CISM International Centre for Mechanical Sciences

Courses and Lectures

Volume 610

For more than 40 years the book series edited by CISM, "International Centre for Mechanical Sciences: Courses and Lectures", has presented groundbreaking developments in mechanics and computational engineering methods. It covers such fields as solid and fluid mechanics, mechanics of materials, micro- and nanomechanics, biomechanics, and mechatronics. The papers are written by international authorities in the field. The books are at graduate level but may include some introductory material.

Kostas Danas · Oscar Lopez-Pamies
Editors

Electro- and Magneto-Mechanics of Soft Solids

Constitutive Modelling, Numerical Implementations, and Instabilities

Springer

Editors
Kostas Danas
Laboratoire de Mécanique des Solides (LMS)
National Center for Scientific Research (CNRS)
École Polytechnique
Palaiseau, France

Oscar Lopez-Pamies
Newmark Laboratory
University of Illinois Urbana-Champaign
Urbana, IL, USA

ISSN 0254-1971 ISSN 2309-3706 (electronic)
CISM International Centre for Mechanical Sciences
ISBN 978-3-031-48350-9 ISBN 978-3-031-48351-6 (eBook)
https://doi.org/10.1007/978-3-031-48351-6

This Springer imprint is published by the registered company Springer Nature Switzerland AG
The registered company address is: Gewerbestrasse 11, 6330 Cham, Switzerland

Preface

In the past ten years, the amount of research on active electro- and magneto-active soft polymer solids has faced an almost exponential increase. These soft composite solids are in most cases polymer composites comprising micron size particles that are electrically conducting or magnetizable. In addition, they are fairly easy and cheap to fabricate and thus are continuously evolving taking different forms and compositions. New theories and numerical methods are developed in parallel to explain their mechanical and multi-physical response observed in various experimental studies. In turn, the understanding of the underlying microscopic mechanisms leading to electro- and magneto-mechanical coupling at the larger scales of application is of primordial importance, since it allows to explain but also guide in a more targeted manner the development and improvement of the coupled response of such soft active solids.

The difficulty and at the same time beauty of these soft coupled polymers lie in the finite strains and nonlinearities exhibited during electric, magnetic and mechanical actuation. Their mechanical softness leads to large deformations even during the application of small magnetic or electric fields and their description becomes inherently complex since one needs to deal with nonlinear elastic and viscoelastic deformations at various scales without neglecting the nonlinearities of the electric and magnetic response. The proposed theories make use of fundamental notions in continuum mechanics and electromagnetic theories developed during the last hundred years. Yet, the induced non-trivial coupling has led to the development of novel material laws, numerical methods and experimental techniques for the proper description and analysis of such composite materials.

As a consequence of their finite strain response, these active materials and structures exhibit more often than not instabilities. Their theoretical and experimental analysis builds upon existing knowledge of large strain instability theories developed in the context of purely mechanical deformations, yet it requires new additional mathematical tools to deal with the coupled response. Those instabilities may be harnessed to obtain exotic structural response and actuation. This rich behavior allows to develop new unprecedented applications in the fields of biomedical engineering, soft robots and locomotion as well as in industrial braking systems of all sorts.

This course was built having in mind all the above-described phenomena and was addressed to advanced students and researchers working in the field of such coupled systems in physics, material science and mechanics. The lectures were given by established scientists in this field with significant contributions to the coupled electro-magneto-mechanical study of such polymers.

The present book contains four individual chapters covering work on the fundamentals (O. Lopez-Pamies) and the modeling (M. Gei) of electroactive solids, the modeling of magneto-active solids (K. Danas) and the analysis of elastic instabilities (Y. Fu).

It is a pleasure to thank the lecturers and the attendees who came from many different countries and actively participated in the course. The efficient help of the administrative staff of CISM is also gratefully acknowledged.

Palaiseau, France Kostas Danas
Urbana, USA Oscar Lopez-Pamies

Contents

The Elastic Dielectric Response of Elastomers Filled with Liquid Inclusions: From Fundamentals to Governing Equations

Oscar Lopez-Pamies

Abstract Over the past decade, soft solids containing electro-and magneto-active *liquid*—as opposed to solid—inclusions have emerged as a new class of smart materials with promising novel electro- and magneto-mechanical properties. In this context, a recent contribution has put forth a continuum theory that describes the macroscopic elastic behavior of elastomers filled with liquid inclusions under quasistatic finite deformations from the bottom up, directly in terms of their microscopic behavior at the length scale of the inclusions. This chapter presents the generalization of that theory to the coupled realm of the elastic dielectric behavior of such an emerging class of filled elastomers when in addition to undergoing quasistatic finite deformations they are subjected to quasistatic electric fields. The chapter starts with the description of the underlying fundamentals in the continuum—*id est*, kinematics, conservation of mass, Maxwell's equations, balance of momenta, and constitutive behavior of both the bulk (the solid elastomer and the liquid inclusions) and the solid/liquid interfaces—and ends with their combination to formulate the resulting governing equations.

1 Introduction

The study of the mechanics of interfaces in the continuum has a long and rich history with origins dating back to the classical works of Young (1805) and Laplace (1806) in the early 1800s on interfaces between fluids and of Gibbs (1878) in the 1870s on the more general case of interfaces between solids and fluids. Despite this early origins, it was only in 1975 that complete descriptions of the kinematics, the concept of interface stress, and the balance of linear and angular momenta of bodies containing material interfaces were properly formulated, alongside their specialization to the basic constitutive case of elastic interfaces (Gurtin and Murdoch (1975a, b)). This

O. Lopez-Pamies (✉)
Department of Civil and Environmental Engineering, University of Illinois, Urbana–Champaign 61801, USA
e-mail: pamies@illinois.edu

K. Danas and O. Lopez-Pamies (eds.), *Electro- and Magneto-Mechanics of Soft Solids*, CISM International Centre for Mechanical Sciences 610,
https://doi.org/10.1007/978-3-031-48351-6_1

advancement in theory, however, was not followed by its exploitation in practice, surely because of the technical difficulties of measuring and tailoring the mechanical properties of interfaces at the time. This changed in the early 2000s, when the appearance of new synthesis and characterization tools reinvigorated the study of interfaces in soft matter.

In this context, elastomers filled with *liquid*—as opposed to solid—inclusions are a recent trend in the soft matter community because—thanks to the behavior of the solid/liquid interfaces—they are capable of exhibiting remarkable mechanical and physical properties; see, e.g., Lopez-Pamies (2014), Style et al. (2015a), Bartlett et al. (2017), Lefèvre and Lopez-Pamies (2017a, b), Lefèvre et al. (2017) and Yun et al. (2019). Indeed, the interfacial mechanics in these soft material systems can be actively tailored to enhance or impede deformability. In particular, while the addition of liquid inclusions should increase the macroscopic deformability of the material, the behavior of the solid/liquid interfaces, if negligible when the inclusions are "large", may counteract the macroscopic properties of the material when the inclusions become sufficiently "small".

As a first step to understand in a precise and quantitative manner this new paradigm, Ghosh and Lopez-Pamies (2022) have recently worked out the governing equations that describe from the bottom up the mechanical response of an elastic solid filled with initially spherical inclusions made of a pressurized elastic fluid when the solid/fluid interfaces are elastic and possess an initial surface tension; see also Ghosh et al. (2023a, b) for applications, Díaz et al. (2023) for the associated mathematical analysis, as well as Style et al. (2015b), Wang and Henann (2016), and Krichen et al. (2019) for earlier preliminary studies.

In this chapter, we generalize the formulation of Ghosh and Lopez-Pamies (2022) to the coupled realm of elastic dielectric behavior. In particular, we work out the governing equations that describe the electromechanical response of an elastic dielectric solid filled with initially spherical inclusions made of a pressurized elastic dielectric fluid when the solid/fluid interfaces feature their own elastic dielectric behavior and possess an initial surface tension.

The organization of the chapter is as follows. In Sect. 2 through Sect. 6, we present separately the relevant basic ingredients of:

- initial configuration and kinematics of the bulk and interfaces,
- conservation of mass,
- Maxwell's equations in the presence of material interfaces,
- balance of linear and angular momenta in the presence of material interfaces, and
- elastic dielectric constitutive behavior of the bulk and interfaces.

The combination of these ingredients leads to the governing equations that describe the elastic dielectric response of elastomers filled with liquid inclusions under finite quasistatic deformations and quasistatic electric fields. We present these in Sect. 7.

2 Initial Configuration and Kinematics

Initial configuration Consider a body made of M liquid inclusions fully embedded in a solid matrix that in its initial configuration occupies the open domain $\Omega_0 \subset \mathbb{R}^3$, with boundary $\partial\Omega_0$ and outward unit normal $\mathbf{N}$. Denote by $\Omega_0^{\mathtt{m}}$ the subdomain occupied by the matrix and by $\Omega_0^{\mathtt{i},j}$ $j = 1, 2, ..., M$ that occupied by the jth inclusion. The inclusions are separated from the matrix by smooth interfaces, denoted by Γ_0^j for the jth inclusion, with unit normal $\widehat{\mathbf{N}}$ pointing outwards from the inclusions towards the matrix, so that $\Omega_0 = \Omega_0^{\mathtt{m}} \bigcup \Gamma_0 \bigcup \Omega_0^{\mathtt{i}}$, where $\Gamma_0 = \bigcup_{j=1}^{M} \Gamma_0^j$ and $\Omega_0^{\mathtt{i}} = \bigcup_{j=1}^{M} \Omega_0^{\mathtt{i},j}$. We identify material points in the body by their initial position vector

$$\mathbf{X} \in \Omega_0$$

and denote by $\theta_0^{\mathtt{i},j}(\mathbf{X})$ and $\theta_0^{\mathtt{i}}(\mathbf{X})$ the characteristic or indicator functions describing the individual and collective spatial locations occupied by the inclusions in Ω_0, that is,

$$\theta_0^{\mathtt{i},j}(\mathbf{X}) = \begin{cases} 1 & \text{if } \mathbf{X} \in \Omega_0^{\mathtt{i},j} \\ 0 & \text{otherwise} \end{cases} \quad j = 1, 2, ..., M \quad \text{and} \quad \theta_0^{\mathtt{i}}(\mathbf{X}) = \sum_{j=1}^{M} \theta_0^{\mathtt{i},j}(\mathbf{X}). \tag{1}$$

As will become apparent below, it is convenient to single out the material points on the interfaces with their own labeling. We write

$$\widehat{\mathbf{X}} = \mathbf{X} \quad \text{when} \quad \mathbf{X} \in \Gamma_0.$$

Figure 1a shows a schematic of the body in its initial configuration with all the pertinent geometric quantities depicted.

Kinematics In response to the externally applied mechanical forces and electric fields described further below, the position vector $\mathbf{X}$ of a material point may occupy a new position $\mathbf{x}$ specified by a continuous,[1] invertible, orientation-preserving mapping $\mathbf{y}$ from Ω_0 to the current configuration $\Omega = \Omega^{\mathtt{m}} \bigcup \Gamma \bigcup \Omega^{\mathtt{i}} \subset \mathbb{R}^3$, termed the *deformation field*. Here, in direct analogy to their initial counterparts, $\Omega^{\mathtt{m}}$, $\Omega^{\mathtt{i}}$, and Γ denote the subdomains occupied by the matrix and the inclusions and the interfaces separating them; by the same token, the notation $\widehat{\mathbf{n}}$, $\theta^{\mathtt{i},j}$, and $\theta^{\mathtt{i}}$ is used to denote the counterparts of $\widehat{\mathbf{N}}$, $\theta_0^{\mathtt{i},j}$, and $\theta_0^{\mathtt{i}}$ in the current configuration. We write

$$\mathbf{x} = \mathbf{y}(\mathbf{X}).$$

Singling out again the material points on the interfaces with their own labeling, we also write

$$\widehat{\mathbf{x}} = \mathbf{y}(\widehat{\mathbf{X}}).$$

[1] The focus here is on liquid inclusions, which naturally exhibit coherent interfaces with the surrounding solid matrix, thus our restriction to continuous deformation fields.

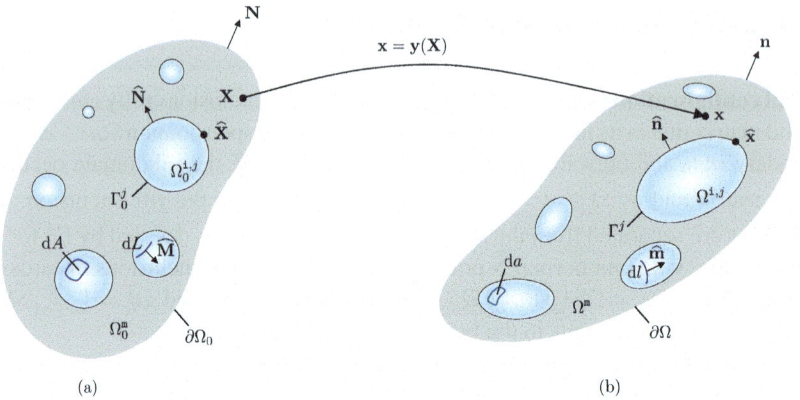

Fig. 1 Schematics of **a** the initial and **b** the current configurations of a body made of a solid matrix filled with liquid inclusions

We denote the deformation gradient at $\mathbf{X} \in \Omega_0$ by

$$\mathbf{F}(\mathbf{X}) = \nabla\mathbf{y}(\mathbf{X}) = \frac{\partial \mathbf{y}}{\partial \mathbf{X}}(\mathbf{X})$$

and the *interface deformation gradient* at $\widehat{\mathbf{X}} \in \Gamma_0$ by

$$\widehat{\mathbf{F}}(\widehat{\mathbf{X}}) = \widehat{\nabla}\mathbf{y}(\widehat{\mathbf{X}}) = \mathbf{F}(\widehat{\mathbf{X}})\widehat{\mathbf{I}}, \tag{2}$$

where $\widehat{\mathbf{I}}$ stands for the projection tensor

$$\widehat{\mathbf{I}} = \mathbf{I} - \widehat{\mathbf{N}} \otimes \widehat{\mathbf{N}}.$$

The notation (2) merits some clarification. Assuming sufficient regularity away from the interfaces, the requirement that the deformation field $\mathbf{y}(\mathbf{X})$ be continuous implies the Hadamard jump condition

$$[\![\mathbf{F}(\widehat{\mathbf{X}})]\!]\widehat{\mathbf{I}} = \mathbf{0} \quad \text{with} \qquad [\![\mathbf{F}(\widehat{\mathbf{X}})]\!] := \mathbf{F}^{\mathtt{i}}(\widehat{\mathbf{X}}) - \mathbf{F}^{\mathtt{m}}(\widehat{\mathbf{X}}), \tag{3}$$

where $\mathbf{F}^{\mathtt{i}}$ ($\mathbf{F}^{\mathtt{m}}$) denotes the limit of $\mathbf{F}$ when approaching Γ_0 from $\Omega_0^{\mathtt{i}}$ ($\Omega_0^{\mathtt{m}}$). Although $\mathbf{F}^{\mathtt{i}} \neq \mathbf{F}^{\mathtt{m}}$ at Γ_0, $\mathbf{F}^{\mathtt{i}}\widehat{\mathbf{I}} = \mathbf{F}^{\mathtt{m}}\widehat{\mathbf{I}}$, and it is for this reason that, with some abuse of notation, we do not include the label '$\mathtt{i}$' or '$\mathtt{m}$' in the right-hand side of (2). In the sequel, at solid/liquid interfaces, we always make use of the convention $[\![\cdot]\!] := (\cdot)^{\mathtt{i}} - (\cdot)^{\mathtt{m}}$ for the jump operator acting on any field.

The interested reader is referred to, for instance, Carmo (2016), Weatherburn (2016), Gurtin et al. (1998) and Javili et al. (2013) for a thorough description of differential operators defined on surfaces embedded in $\mathbb{R}^3$ and of the kinematics of interfaces. For our purposes here, it suffices to make explicit mention of some of the properties of the interface deformation gradient (2).

In direct analogy to the transformation rules for material line elements $\mathrm{d}\mathbf{X}$ in the bulk, material line elements $\mathrm{d}\widehat{\mathbf{X}}$ on the interfaces transform according to the rules

$$\mathrm{d}\widehat{\mathbf{x}} = \widehat{\mathbf{F}}\mathrm{d}\widehat{\mathbf{X}} \quad \text{and} \quad \mathrm{d}\widehat{\mathbf{X}} = \widehat{\mathbf{F}}^{-1}\mathrm{d}\widehat{\mathbf{x}}.$$

Owing to its rank deficiency, the inverse $\widehat{\mathbf{F}}^{-1}$ of the interface gradient deformation $\widehat{\mathbf{F}}$ is defined implicitly by the relations

$$\widehat{\mathbf{F}}^{-1}\widehat{\mathbf{F}} = \widehat{\mathbf{I}} \quad \text{and} \quad \widehat{\mathbf{F}}\widehat{\mathbf{F}}^{-1} = \widehat{\mathbf{i}}.$$

where

$$\widehat{\mathbf{i}} = \mathbf{I} - \widehat{\mathbf{n}} \otimes \widehat{\mathbf{n}} \quad \text{with} \quad \widehat{\mathbf{n}} = \frac{1}{|J\mathbf{F}^{-T}\widehat{\mathbf{N}}|} J\mathbf{F}^{-T}\widehat{\mathbf{N}}.$$

That is

$$\widehat{\mathbf{F}}^{-1} = \mathbf{F}^{-1}\widehat{\mathbf{i}}.$$

In these last expressions, we have made use of the standard notation $J = \det \mathbf{F}$ for the determinant of the deformation gradient $\mathbf{F}$ and exploited the facts that $J^{\mathtt{i}}\mathbf{F}^{\mathtt{i}^{-T}}\widehat{\mathbf{N}} = J^{\mathtt{m}}\mathbf{F}^{\mathtt{m}-T}\widehat{\mathbf{N}}$ and $\mathbf{F}^{\mathtt{i}^{-1}}\widehat{\mathbf{i}} = \mathbf{F}^{\mathtt{m}-1}\widehat{\mathbf{i}}$, thanks to (3), to simply write, with the same abuse of notation as in (2), $J\mathbf{F}^{-T}\widehat{\mathbf{N}}$ and $\mathbf{F}^{-1}\widehat{\mathbf{i}}$ without the label '$\mathtt{i}$' or '$\mathtt{m}$'.

Furthermore, the area $\mathrm{d}A$ of material surface elements $\widehat{\mathbf{N}}\mathrm{d}A$ on the interfaces transforms according to the rule

$$\mathrm{d}a = \widehat{J}\mathrm{d}A \quad \text{with} \quad \widehat{J} = |J\mathbf{F}^{-T}\widehat{\mathbf{N}}|.$$

This transformation rule also serves to define the interface determinant operator $\widehat{\det}\,\widehat{\mathbf{F}} = \widehat{J}$.

Finally, we note that material curve elements $\widehat{\mathbf{M}}\,\mathrm{d}L$ on the interfaces transform according to the rule

$$\widehat{\mathbf{m}}\,\mathrm{d}l = \widehat{J}\widehat{\mathbf{F}}^{-T}\widehat{\mathbf{M}}\,\mathrm{d}L,$$

where $\widehat{\mathbf{M}}$ is a unit vector that is tangential to Γ_0 and normal to the curve $\mathrm{d}L$. Figure 1b provides a schematic of the body in its current configuration with all the above geometric quantities depicted.

3 Conservation of Mass

In any given subdomain of the current configuration $\mathcal{D} \subset \Omega$ we consider the existence of a *mass density*

$$\rho(\mathbf{x}) \geq 0, \quad \mathbf{x} \in \mathcal{D}.$$

Integral Form Conservation of mass then reads

$$\int_{\mathcal{D}} \rho \, \mathrm{d}\mathbf{x} = \int_{\mathcal{D}_0} \rho_0 \, \mathrm{d}\mathbf{X}, \tag{4}$$

where $\rho_0(\mathbf{X})$ is the mass density in the initial configuration.

Localized Form It follows from the integral form (4) that

$$\int_{\mathcal{D}_0} J\rho \, \mathrm{d}\mathbf{X} = \int_{\mathcal{D}_0} \rho_0 \, \mathrm{d}\mathbf{X}$$

and hence that conservation of mass can be rewritten in the localized form

$$\rho = J^{-1}\rho_0, \quad \mathbf{x} \in \Omega \setminus \Gamma. \tag{5}$$

4 Maxwell's Equations in the Presence of Material Interfaces

4.1 Bulk and Interface Charges, Electric Fields, and Electric Displacements

In any given subdomain of the current configuration $\mathcal{D} \subset \Omega$, with boundary $\partial\mathcal{D}$ and outward unit normal $\widetilde{\mathbf{n}}$, we consider the presence of a *space charge* density per unit current volume

$$q(\mathbf{x}), \quad \mathbf{x} \in \mathcal{D},$$

an *electric displacement*

$$\mathbf{d}(\mathbf{x}), \quad \mathbf{x} \in \mathcal{D},$$

an *electric field*

$$\mathbf{e}(\mathbf{x}), \quad \mathbf{x} \in \mathcal{D},$$

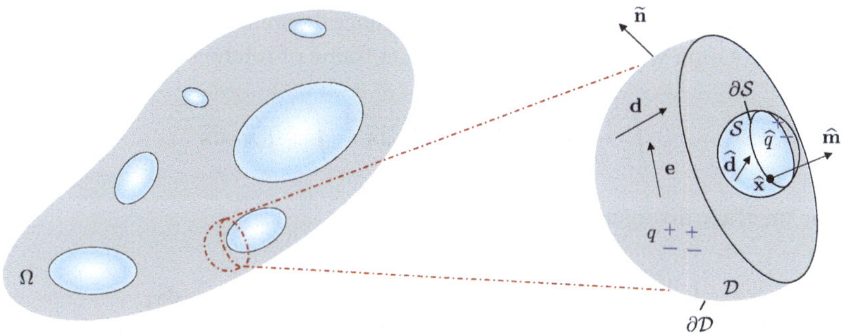

Fig. 2 Schematic of a subdomain of the current configuration $\mathcal{D} \subset \Omega$, with boundary $\partial\mathcal{D}$ and outward unit normal $\widetilde{\mathbf{n}}$, indicating the space charge $q(\mathbf{x})$, the electric displacement $\mathbf{d}(\mathbf{x})$, the electric field $\mathbf{e}(\mathbf{x})$, the interface charge $\widehat{q}(\widehat{\mathbf{x}})$, and the interface electric displacement $\widehat{\mathbf{d}}(\widehat{\mathbf{x}})$

an *interface charge* density per unit current area

$$\widehat{q}(\widehat{\mathbf{x}}), \quad \widehat{\mathbf{x}} \in \mathcal{S},$$

and an *interface electric displacement*

$$\widehat{\mathbf{d}}(\widehat{\mathbf{x}}), \quad \widehat{\mathbf{x}} \in \mathcal{S},$$

where $\mathcal{S} \subset \Gamma$, with boundary $\partial\mathcal{S}$, stands for any subsurfaces of the interfaces that the subdomain $\mathcal{D}$ may contain. The first four of these quantities are standard. The fifth one accounts for the possibility of an additional polarization mechanism at the matrix/inclusions interfaces. Within such a class of interface electric displacements, we shall restrict attention to *tangential* electric displacements in the sense that

$$\widehat{\mathbf{i}}\widehat{\mathbf{d}} = \widehat{\mathbf{d}}. \tag{6}$$

Figure 2 shows a schematic of a generic subdomain $\mathcal{D}$ with all five types of bulk and interface quantities depicted.

4.2 *Gauss's Law*

Integral Form In view of the presence of bulk and interface charges and the interface electric displacement, the integral form of Gauss's law reads

$$\int_{\partial\mathcal{D}} \mathbf{d} \cdot \widetilde{\mathbf{n}} \, \mathrm{d}\mathbf{x} + \int_{\partial\mathcal{S}} \widehat{\mathbf{d}} \cdot \widehat{\mathbf{m}} \, \mathrm{d}\widehat{\mathbf{x}} = \int_{\mathcal{D}} q \, \mathrm{d}\mathbf{x} + \int_{\mathcal{S}} \widehat{q} \, \mathrm{d}\widehat{\mathbf{x}}. \tag{7}$$

Localized Form Making use of the bulk divergence theorem (written here, for clarity, in indicial notation with respect to a Cartesian frame of reference)

$$\int_{\mathcal{D}} \frac{\partial(\cdot)}{\partial x_k}\,\mathrm{d}\mathbf{x} = \int_{\partial\mathcal{D}} (\cdot)\widetilde{n}_k\,\mathrm{d}\mathbf{x} + \int_{\mathcal{S}} [\![\cdot]\!]\,\widehat{n}_k\,\mathrm{d}\widehat{\mathbf{x}} \tag{8}$$

and the interface divergence theorem

$$\int_{\mathcal{S}} \frac{\partial(\cdot)}{\partial x_l}\widehat{\imath}_{kl}\,\mathrm{d}\widehat{\mathbf{x}} = \int_{\partial\mathcal{S}} (\cdot)\widehat{m}_k\,\mathrm{d}\widehat{\mathbf{x}} + \int_{\mathcal{S}} \frac{\partial\widehat{n}_p}{\partial x_q}\widehat{\imath}_{pq}(\cdot)\widehat{n}_k\,\mathrm{d}\widehat{\mathbf{x}}, \tag{9}$$

together with the fact that the interface electric displacement $\widehat{\mathbf{d}}$ is a tangential vector—so that $\widehat{\mathbf{d}}\cdot\widehat{\mathbf{n}} = 0$ as a consequence of (6)—Gauss's law (7) can be rewritten in the localized or differential form

$$\begin{cases} \operatorname{div}\mathbf{d} = q, & \mathbf{x}\in\Omega\setminus\Gamma \\ \widehat{\operatorname{div}}\,\widehat{\mathbf{d}} - [\![\mathbf{d}]\!]\cdot\widehat{\mathbf{n}} = \widehat{q}, & \mathbf{x}\in\Gamma \\ \operatorname{div}\mathbf{d} = 0, & \mathbf{x}\in\mathbb{R}^3\setminus\Omega \\ [\![\mathbf{d}]\!]\cdot\mathbf{n} = 0, & \mathbf{x}\in\partial\Omega \end{cases}. \tag{10}$$

In these expressions, div is the standard divergence operator in the bulk and $\widehat{\operatorname{div}}$ stands for the interface divergence operator, that is, in indicial notation

$$\operatorname{div}\mathbf{d} = \frac{\partial d_k}{\partial x_k} \quad \text{and} \quad \widehat{\operatorname{div}}\,\widehat{\mathbf{d}} = \frac{\partial d_k}{\partial x_l}\widehat{\imath}_{kl}.$$

Remark 1.1 Note that Eq. (10) make it explicit that Gauss's law applies in the entirety of space, and hence also outside the body.

Lagrangian Localized Form Much like its integral counterpart (7), Gauss's law (10) is in Eulerian (spatial) form. For computational purposes, we shall find it more convenient to deal with it in its Lagrangian (material) form

$$\begin{cases} \operatorname{Div}\mathbf{D} = Q, & \mathbf{X}\in\Omega_0\setminus\Gamma_0 \\ \widehat{\operatorname{Div}}\,\widehat{\mathbf{D}} - [\![\mathbf{D}]\!]\cdot\widehat{\mathbf{N}} = \widehat{Q}, & \mathbf{X}\in\Gamma_0 \\ \operatorname{Div}\mathbf{D} = 0, & \mathbf{X}\in\mathbb{R}^3\setminus\Omega \\ [\![\mathbf{D}]\!]\cdot\mathbf{N} = 0, & \mathbf{X}\in\partial\Omega_0 \end{cases}, \tag{11}$$

where $Q = Jq$ is the *space charge* density per unit initial volume, $\mathbf{D} = J\mathbf{F}^{-1}\mathbf{d}$ is the *Lagrangian electric displacement*, while $\widehat{Q} = \widehat{Jq}$ stands for the *interface charge* density per unit initial area, and $\widehat{\mathbf{D}} = \widehat{J\mathbf{F}^{-1}}\widehat{\mathbf{d}}$ stands for the *Lagrangian interface electric displacement*.

The notation utilized in Eq. (11) for the bulk and interface divergence operators in the initial configuration is entirely analogous to that employed in (10) in the current configuration: $\text{Div}\,\mathbf{D} = \text{tr}\,\nabla\mathbf{D} = \nabla\mathbf{D}\cdot\mathbf{I}$ and $\widehat{\text{Div}}\,\widehat{\mathbf{D}} = \text{tr}\,\widehat{\nabla}\widehat{\mathbf{D}} = \nabla\widehat{\mathbf{D}}\cdot\widehat{\mathbf{I}}$. A derivation of (11) starting from (10) is provided in Appendix A.

4.3 Faraday's Law

Integral Form In the absence of magnetic fields, electric currents, and time dependence, when Ampère's law and Gauss's law for magnetism are trivially satisfied, the integral form of Faraday's law reads

$$\int_{\partial\Sigma} \mathbf{e}\cdot \mathrm{d}\mathbf{x} = 0, \tag{12}$$

where Σ is any given open surface, with unit normal $\widetilde{\mathbf{n}}$, in the current configuration Ω and $\partial\Sigma$ denotes its boundary, a closed curve oriented in the usual sense with respect to $\widetilde{\mathbf{n}}$; see Fig. 3.

Localized Form Making use of Stokes's theorem

$$\int_{\Sigma} (\text{curl}\,\mathbf{e})\cdot\widetilde{\mathbf{n}}\,\mathrm{d}\mathbf{x} = \int_{\partial\Sigma} \mathbf{e}\cdot\mathrm{d}\mathbf{x} + \int_{\partial\mathcal{S}} [\![\mathbf{e}]\!]\cdot\mathrm{d}\mathbf{x},$$

Faraday's law (12) can be rewritten in the localized form

$$\begin{cases} \text{curl}\,\mathbf{e} = \mathbf{0}, & \mathbf{x}\in\Omega\setminus\Gamma \\ \widehat{\mathbf{i}}\,[\![\mathbf{e}]\!] = \mathbf{0}, & \mathbf{x}\in\Gamma \\ \text{curl}\,\mathbf{e} = \mathbf{0}, & \mathbf{x}\in\mathbb{R}^3\setminus\Omega \\ (\mathbf{I}-\mathbf{n}\otimes\mathbf{n})\,[\![\mathbf{e}]\!] = \mathbf{0}, & \mathbf{x}\in\partial\Omega \end{cases} \tag{13}$$

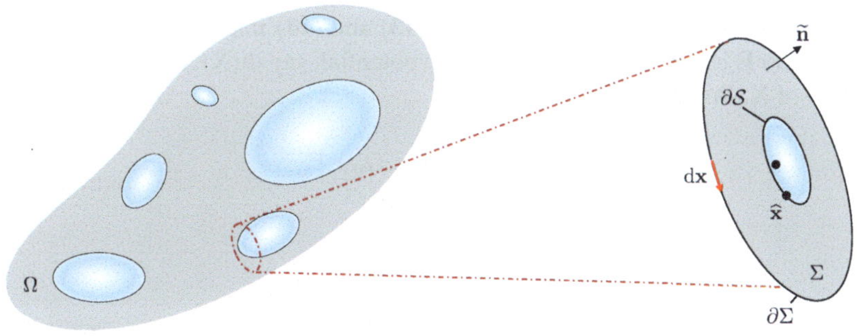

Fig. 3 Schematic of an open surface Σ, cutting through a liquid inclusion, in the current configuration Ω indicating its unit normal $\widetilde{\mathbf{n}}$ and boundary $\partial\Sigma$

In these expressions, curl is the standard curl operator in the bulk, that is, in indicial notation

$$(\mathrm{curl}\,\mathbf{e})_i = \varepsilon_{ijk}\frac{\partial e_k}{\partial x_j}.$$

Remark 1.2 Analogous to (10), Eq. (13) make it explicit that Faraday's law applies in the entirety of space, and hence also outside the body.

Remark 1.3 Equations (13) imply that the electric field **e** is the gradient of an electric potential, say $\phi(\mathbf{x})$, that is a continuous function of **x**. Precisely,

$$\begin{cases} \mathbf{e} = \nabla_{\mathbf{x}}\phi = \dfrac{\partial\phi}{\partial\mathbf{x}}, & \mathbf{x} \in \Omega \setminus \Gamma \\ [\![\phi]\!] = 0, & \mathbf{x} \in \Gamma \\ \mathbf{e} = \nabla_{\mathbf{x}}\phi = \dfrac{\partial\phi}{\partial\mathbf{x}}, & \mathbf{x} \in \mathbb{R}^3 \setminus \Omega \\ [\![\phi]\!] = 0, & \mathbf{x} \in \partial\Omega \end{cases}$$

Lagrangian Localized Form For computational purposes, as already noted for Gauss's law above, we shall find it more convenient not to deal with the Eulerian form (13) of Faraday's law, but to do so with its Lagrangian form. This reads

$$\begin{cases} \mathrm{Curl}\,\mathbf{E} = \mathbf{0}, & \mathbf{X} \in \Omega_0 \setminus \Gamma_0 \\ \widehat{\mathbf{I}}\,[\![\mathbf{E}]\!] = \mathbf{0}, & \mathbf{X} \in \Gamma_0 \\ \mathrm{Curl}\,\mathbf{E} = \mathbf{0}, & \mathbf{X} \in \mathbb{R}^3 \setminus \Omega_0 \\ (\mathbf{I} - \mathbf{N} \otimes \mathbf{N})\,[\![\mathbf{E}]\!] = \mathbf{0}, & \mathbf{X} \in \partial\Omega_0 \end{cases}, \tag{14}$$

where $\mathbf{E} = \mathbf{F}^T\mathbf{e}$ is the *Lagrangian electric field* and the curl operator is entirely analogous to that employed in (13) in the current configuration: $\mathrm{Curl}\,\mathbf{E} = \nabla \wedge \mathbf{E}$. A derivation of (14) starting from (12) is provided in Appendix B.

Remark 1.4 Of course, exactly like Eqs. (13) and (14) imply that the Lagrangian electric field **E** is the gradient of an electric potential, say $\Phi(\mathbf{X})$, that is a continuous function of **X**. Precisely,

$$\begin{cases} \mathbf{E} = \nabla\Phi, & \mathbf{X} \in \Omega_0 \setminus \Gamma_0 \\ [\![\Phi]\!] = 0, & \mathbf{X} \in \Gamma_0 \\ \mathbf{E} = \nabla\Phi, & \mathbf{X} \in \mathbb{R}^3 \setminus \Omega_0 \\ [\![\Phi]\!] = 0, & \mathbf{X} \in \partial\Omega_0 \end{cases}$$

Clearly, $\Phi(\mathbf{X}) = \phi(\mathbf{y}(\mathbf{X}))$.

5 Balance of Momenta in the Presence of Material Interfaces

5.1 Bulk and Interface Electric and Mechanical Forces

In any given subdomain of the current configuration $\mathcal{D} \subset \Omega$, with boundary $\partial\mathcal{D}$ and outward unit normal $\widetilde{\mathbf{n}}$, the presence of the space charge density q, electric field $\mathbf{e}$, and electric displacement $\mathbf{d}$ described in the preceding section generates an *electric surface force* per unit current area, or electric surface traction, given by

$$\mathbf{t}_{\mathsf{e}}(\mathbf{x}) = \underbrace{\left[\mathbf{e} \otimes \mathbf{d} - \frac{1}{2}\varepsilon_0(\mathbf{e} \cdot \mathbf{e})\mathbf{I}\right]}_{\mathbf{T}_{\mathsf{e}}} \widetilde{\mathbf{n}}, \quad \mathbf{x} \in \partial\mathcal{D}, \tag{15}$$

where ε_0 is the permittivity of vacuum and $\mathbf{T}_{\mathsf{e}}$ is the so-called Maxwell stress.

Remark 1.5 Making use of the bulk divergence theorem (8), the electric surface traction (15) implies the presence of an *electric body force* per unit current volume, which reads

$$\mathbf{b}_{\mathsf{e}}(\mathbf{x}) = \operatorname{div} \mathbf{T}_{\mathsf{e}} = q\mathbf{e} + (\nabla_{\mathbf{x}}\mathbf{e})\,(\mathbf{d} - \varepsilon_0\mathbf{e})\,, \quad \mathbf{x} \in \mathcal{D};$$

see, e.g., Eq. (7.38) in the review by Pao (1978).

Remark 1.6 In direct analogy to (15), the interface charge density $\widehat{q}$ and interface electric displacement $\widehat{\mathbf{d}}$ described in the preceding section generate an *electric interface force* per unit current length, or electric interface traction, say $\widehat{\mathbf{t}}_{\mathsf{e}}(\widehat{\mathbf{x}})$, $\widehat{\mathbf{x}} \in \partial\mathcal{S}$. In this work, we take such an electric interface force to be negligible compared to the rest of forces.

In addition to the electric surface force (15), we consider that there may be three different types of mechanical forces present in any given subdomain $\mathcal{D} \subset \Omega$, to wit, a *mechanical body force* per unit current volume

$$\mathbf{b}_{\mathsf{m}}(\mathbf{x}), \quad \mathbf{x} \in \mathcal{D},$$

a *mechanical surface force* per unit current area, or mechanical surface traction,

$$\mathbf{t}_{\mathsf{m}}(\mathbf{x}, \widetilde{\mathbf{n}}), \quad \mathbf{x} \in \partial\mathcal{D},$$

and a *mechanical interface force* per unit current length, or mechanical interface traction,

$$\widehat{\mathbf{t}}_{\mathsf{m}}(\widehat{\mathbf{x}}, \widehat{\mathbf{m}}), \quad \widehat{\mathbf{x}} \in \partial\mathcal{S}.$$

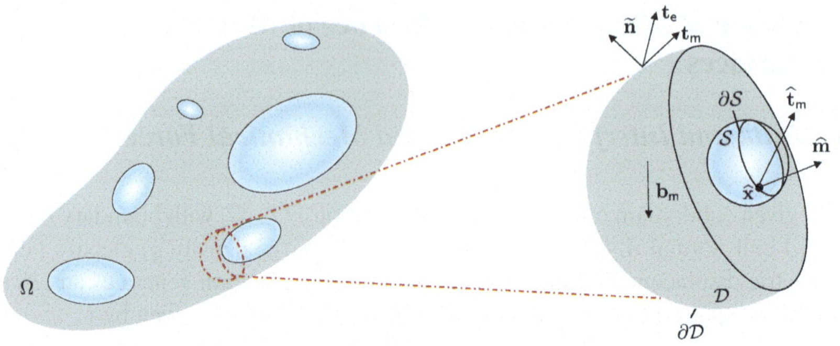

Fig. 4 Schematic of a subdomain of the current configuration $\mathcal{D} \subset \Omega$, with boundary $\partial\mathcal{D}$ and outward unit normal $\widetilde{\mathbf{n}}$, indicating the electric surface force $\mathbf{t}_e(\mathbf{x})$, mechanical body force $\mathbf{b}_m(\mathbf{x})$, mechanical surface force $\mathbf{t}_m(\mathbf{x}, \widetilde{\mathbf{n}})$, and mechanical interface force $\widehat{\mathbf{t}}_m(\widehat{\mathbf{x}}, \widehat{\mathbf{m}})$ that is subjected to

Recall that $\partial\mathcal{S}$ stands for the boundary of any subsurfaces of the interfaces $\mathcal{S} \subset \Gamma$ that the subdomain $\mathcal{D}$ may contain. The first two of these three mechanical forces are standard. The third one accounts for the possibility of additional forces at the matrix/inclusions interfaces, such as, for instance, surface tension and Marangoni forces; see, e.g., Popinet (2018) and references therein. Within such a class of interface forces, in analogy to (6), we shall restrict attention to *tangential* forces in the sense that

$$\widehat{\mathbf{i}}\widehat{\mathbf{t}}_m = \widehat{\mathbf{t}}_m. \tag{16}$$

Note that Cauchy's fundamental postulate has been tacitly assumed to apply, thus the dependencies of the surface traction $\mathbf{t}_m$ on $\widetilde{\mathbf{n}}$ and of the interfacial traction $\widehat{\mathbf{t}}_m$ on $\widehat{\mathbf{m}}$, which, again, stands for the outward unit normal to $\partial\mathcal{S}$. Figure 4 shows a schematic of a generic subdomain $\mathcal{D}$ with all four types of forces depicted.

5.2 *Balance of Linear Momentum*

Integral Form Absent inertia, granted the above-described types of electric and mechanical forces, balance of linear momentum reads

$$\int_{\partial\mathcal{D}} \mathbf{t}_e(\mathbf{x})\, d\mathbf{x} + \int_{\mathcal{D}} \mathbf{b}_m(\mathbf{x})\, d\mathbf{x} + \int_{\partial\mathcal{D}} \mathbf{t}_m(\mathbf{x}, \widetilde{\mathbf{n}})\, d\mathbf{x} + \int_{\partial\mathcal{S}} \widehat{\mathbf{t}}_m(\widehat{\mathbf{x}}, \widehat{\mathbf{m}})\, d\widehat{\mathbf{x}} = \mathbf{0}. \tag{17}$$

Assuming that $\mathbf{t}_m$ and $\widehat{\mathbf{t}}_m$ are continuous in $\partial\mathcal{D} \setminus \partial\mathcal{S}$ and $\partial\mathcal{S}$, respectively, if follows from (17) that

$$\mathbf{t}_m(\mathbf{x}, \widetilde{\mathbf{n}}) = \mathbf{T}_m(\mathbf{x})\widetilde{\mathbf{n}}, \quad \mathbf{x} \in \mathcal{D} \setminus \mathcal{S}, \qquad \widehat{\mathbf{t}}_m(\widehat{\mathbf{x}}, \widehat{\mathbf{m}}) = \widehat{\mathbf{T}}_m(\widehat{\mathbf{x}})\widehat{\mathbf{m}}, \quad \widehat{\mathbf{x}} \in \mathcal{S}, \tag{18}$$

where $\mathbf{T}_{\mathrm{m}}$ is the standard Cauchy stress tensor in the bulk (resulting from mechanical forces) while $\widehat{\mathbf{T}}_{\mathrm{m}}$ is the *interface Cauchy stress* tensor. The former is continuous in $\mathcal{D} \setminus \mathcal{S}$ but may have a jump at $\mathcal{S}$, while the latter is continuous on $\mathcal{S}$ and, by virtue of (16), is a *tangential* tensor in the sense that $\widehat{\mathbf{i}}\widehat{\mathbf{T}}_{\mathrm{m}}\widehat{\mathbf{i}} = \widehat{\mathbf{T}}_{\mathrm{m}}$.

Localized Form Making use of relations (18), the bulk divergence theorem (8), the interface divergence theorem (9), and the fact that $\widehat{\mathbf{T}}$ is a *superficial* tensor in the sense that $\widehat{\mathbf{T}}_{\mathrm{m}}\widehat{\mathbf{i}} = \widehat{\mathbf{T}}_{\mathrm{m}}$, the integral form (17) of the balance of linear momentum can be rewritten as

$$\int_{\mathcal{D}} \operatorname{div} \mathbf{T}_{\mathrm{e}} \, \mathrm{d}\mathbf{x} - \int_{\mathcal{S}} [\![\mathbf{T}_{\mathrm{e}}]\!] \, \widehat{\mathbf{n}} \, \mathrm{d}\widehat{\mathbf{x}} + \int_{\mathcal{D}} \mathbf{b}_{\mathrm{m}}(\mathbf{x}) \, \mathrm{d}\mathbf{x} + \int_{\mathcal{D}} \operatorname{div} \mathbf{T}_{\mathrm{m}} \, \mathrm{d}\mathbf{x} - \int_{\mathcal{S}} [\![\mathbf{T}_{\mathrm{m}}]\!] \, \widehat{\mathbf{n}} \, \mathrm{d}\widehat{\mathbf{x}} + \int_{\mathcal{S}} \widehat{\operatorname{div}} \, \widehat{\mathbf{T}}_{\mathrm{m}} \, \mathrm{d}\widehat{\mathbf{x}} = \mathbf{0},$$

from which one can readily determine the localized form

$$\begin{cases} \operatorname{div} \mathbf{T} + \mathbf{b} = \mathbf{0}, & \mathbf{x} \in \Omega \setminus \Gamma \\ \widehat{\operatorname{div}} \, \widehat{\mathbf{T}} - [\![\mathbf{T}]\!] \, \widehat{\mathbf{n}} = \mathbf{0}, & \mathbf{x} \in \Gamma \end{cases} \tag{19}$$

in terms of the *total Cauchy stress tensor*

$$\mathbf{T} = \mathbf{T}_{\mathrm{m}} + \mathbf{T}_{\mathrm{e}} \tag{20}$$

in the bulk.

In these last expressions, for notational simplicity, we have dropped the subscript "m" in

$$\mathbf{b}_{\mathrm{m}} \mapsto \mathbf{b} \quad \text{and} \quad \widehat{\mathbf{T}}_{\mathrm{m}} \mapsto \widehat{\mathbf{T}}$$

since there is no longer risk of confusion. We also recall that div is the standard divergence operator in the bulk, while $\widehat{\operatorname{div}}$ stands for the interface divergence operator, namely, in indicial notation

$$(\operatorname{div} \mathbf{T})_i = \frac{\partial T_{ij}}{\partial x_j} \quad \text{and} \quad \left(\widehat{\operatorname{div}} \, \widehat{\mathbf{T}}\right)_i = \frac{\partial \widehat{T}_{ij}}{\partial x_k} \widehat{i}_{jk}.$$

Lagrangian Localized Form For computational purposes, once more, as already noted for the Maxwell's equations above, we favor dealing with the Lagrangian form of the equations of balance of linear momentum. Those read

$$\begin{cases} \operatorname{Div} \mathbf{S} + \mathbf{B} = \mathbf{0}, & \mathbf{X} \in \Omega_0 \setminus \Gamma_0 \\ \widehat{\operatorname{Div}} \, \widehat{\mathbf{S}} - [\![\mathbf{S}]\!] \, \widehat{\mathbf{N}} = \mathbf{0}, & \mathbf{X} \in \Gamma_0 \end{cases}, \tag{21}$$

where $\mathbf{S} = J\mathbf{T}\mathbf{F}^{-T}$ is the *total first Piola-Kirchhoff stress tensor* in the bulk, $\mathbf{B} = J\mathbf{b}$ is the *mechanical body force* per unit initial volume, and $\widehat{\mathbf{S}} = \widehat{J}\widehat{\mathbf{T}}\widehat{\mathbf{F}}^{-T}$ stands for the *interface first Piola-Kirchhoff stress* tensor.

The notation utilized in (21) for the bulk divergence and interface divergence operators in the initial configuration is entirely analogous to that employed in (19) in the current configuration: $\mathrm{Div}\,\mathbf{S} = \nabla\mathbf{S}\cdot\mathbf{I}$ and $\widehat{\mathrm{Div}}\,\widehat{\mathbf{S}} = \nabla\widehat{\mathbf{S}}\cdot\widehat{\mathbf{I}}$. A derivation of (21) starting from (19) follows *mutatis mutandis* from the derivation given in Appendix A of (Ghosh and Lopez-Pamies (2022)) for the case when only mechanical forces are present.

Remark 1.7 It follows from the connection $\widehat{\mathbf{F}}^{-T} = \widehat{\mathbf{i}}\widehat{\mathbf{F}}^{-T}\widehat{\mathbf{I}}$ that the interface first Piola-Kirchhoff stress $\widehat{\mathbf{S}}$ is a *superficial* tensor in the sense that $\widehat{\mathbf{S}}\widehat{\mathbf{I}} = \widehat{\mathbf{S}}$. Contrary to $\widehat{\mathbf{T}}$, however, $\widehat{\mathbf{S}}$ is *not* a *tangential* tensor since, in general, $\widehat{\mathbf{I}}\widehat{\mathbf{S}}\widehat{\mathbf{I}} \neq \widehat{\mathbf{S}}$.

5.3 Balance of Angular Momentum

Integral form In turn, absent inertia and granted the above-described types of electric and mechanical forces, balance of angular momentum reads

$$\int_{\partial\mathcal{D}} \mathbf{x}\wedge\mathbf{t}_{\mathrm{e}}(\mathbf{x})\,\mathrm{d}\mathbf{x} + \int_{\mathcal{D}} \mathbf{x}\wedge\mathbf{b}_{\mathrm{m}}(\mathbf{x})\,\mathrm{d}\mathbf{x} + \int_{\partial\mathcal{D}} \mathbf{x}\wedge\mathbf{t}_{\mathrm{m}}(\mathbf{x},\widetilde{\mathbf{n}})\,\mathrm{d}\mathbf{x} + \int_{\partial\mathcal{S}} \mathbf{x}\wedge\widehat{\mathbf{t}}_{\mathrm{m}}(\widehat{\mathbf{x}},\widehat{\mathbf{m}})\,\mathrm{d}\widehat{\mathbf{x}} = \mathbf{0}. \tag{22}$$

Localized form A standard calculation (see, e.g., Sect. 3.3.2 in the monograph by Ogden (1997)) shows that the integral form (22) of the balance of angular momentum can be written in the simple localized form

$$\begin{cases} \mathbf{T}^T = \mathbf{T}, & \mathbf{x}\in\Omega\setminus\Gamma \\ \widehat{\mathbf{T}}^T = \widehat{\mathbf{T}}, & \mathbf{x}\in\Gamma \end{cases} \tag{23}$$

in terms of the total Cauchy stress tensor (20) in the bulk and the interface Cauchy stress tensor $\widehat{\mathbf{T}}$, where, for notational simplicity, we have again dropped the subscript "m": $\widehat{\mathbf{T}}_{\mathrm{m}} \mapsto \widehat{\mathbf{T}}$.

Lagrangian localized form Given the definitions $\mathbf{S} = J\mathbf{T}\mathbf{F}^{-T}$ and $\widehat{\mathbf{S}} = \widehat{J}\widehat{\mathbf{T}}\widehat{\mathbf{F}}^{-T}$ of the total bulk and interface first Piola-Kirchhoff stress tensors, it is a simple matter to deduce from the balance of angular momentum (23) in Eulerian form that the balance of angular momentum in Lagrangian form is given by

$$\begin{cases} \mathbf{S}\mathbf{F}^T = \mathbf{F}\mathbf{S}^T, & \mathbf{X}\in\Omega_0\setminus\Gamma_0 \\ \widehat{\mathbf{S}}\widehat{\mathbf{F}}^T = \widehat{\mathbf{F}}\widehat{\mathbf{S}}^T, & \mathbf{X}\in\Gamma_0 \end{cases}. \tag{24}$$

6 Constitutive Behavior

For a given initial configuration Ω_0 of the body, given initial mass density ρ_0, given bulk and interface charges Q and $\widehat{Q}$, and given mechanical body force $\mathbf{B}$, mass conservation (5), the Maxwell's Eqs. (11) and (14), and the balance of linear and angular momenta (21) and (24) are *coupled* equations for the deformation field $\mathbf{y}$, the mass density ρ, the Lagrangian electric displacement $\mathbf{D}$, the Lagrangian electric field $\mathbf{E}$, the Lagrangian interface electric displacement $\widehat{\mathbf{D}}$, the total first Piola-Kirchhoff stress tensor $\mathbf{S}$, and the interface first Piola-Kirchhoff stress tensor $\widehat{\mathbf{S}}$ that apply *generally*.

The next step in the formulation of a mathematically closed system of governing equations is to describe the intrinsic electromechanical properties of the materials that the body is made of, precisely, the constitutive behavior of: the solid that the matrix occupying the subdomain $\Omega_0^{\rm m}$ is made of, the liquid that the inclusions occupying the subdomain $\Omega_0^{\rm i}$ are made of, and the solid/liquid interfaces Γ_0.

6.1 *Constitutive Behavior of the Bulk: The Solid Matrix and the Liquid Inclusions*

Constitutive Behavior of the Solid Matrix The focus of this work is on material systems wherein the underlying solid matrix is an elastomer. Accordingly, neglecting dissipative phenomena, we model the matrix as an elastic dielectric solid. Precisely, making use of the formulation introduced by Dorfmann and Ogden (2005), we find it convenient to characterize the electromechanical behavior of the solid matrix in a Lagrangian formulation by a total free energy (per unit initial volume)

$$W_{\rm m} = W_{\rm m}(\mathbf{F}, \mathbf{E}),$$

that is an objective function of the deformation gradient tensor $\mathbf{F}$ and an objective and even function of the Lagrangian electric field $\mathbf{E}$, so that

$$W_{\rm m}(\mathbf{QF}, \mathbf{E}) = W_{\rm m}(\mathbf{F}, \mathbf{E}) \quad \text{and} \quad W_{\rm m}(\mathbf{F}, -\mathbf{E}) = W_{\rm m}(\mathbf{F}, \mathbf{E}) \quad \forall \mathbf{Q} \in SO(3)$$

and arbitrary $\mathbf{F}$ and $\mathbf{E}$.

In the sequel, for clarity of presentation, we will restrict attention to the basic case of an ideal elastic dielectric for which the free energy reads

$$W_{\rm m}(\mathbf{F}, \mathbf{E}) = \frac{\mu_{\rm m}}{2}\left[\mathbf{F}\cdot\mathbf{F} - 3\right] - \mu_{\rm m} \ln J + \frac{\Lambda_{\rm m}}{2}(J-1)^2 - \frac{\varepsilon_{\rm m}}{2} J\mathbf{F}^{-T}\mathbf{E}\cdot\mathbf{F}^{-T}\mathbf{E}. \quad (25)$$

In this constitutive prescription, the material constants $\mu_{\mathtt{m}} > 0$, $\Lambda_{\mathtt{m}} > 0$, $\varepsilon_{\mathtt{m}} \geq \varepsilon_0$ stand for the initial Lamé constants and the initial permittivity of the elastomer under consideration.

Remark 1.8 Here, it is important to emphasize that the use of a free energy of the form (25) implies that, in its initial configuration, the elastomeric matrix is stress and polarization free. In other words, we are assuming that there are *no* residual stress and *no* residual polarization in the elastomeric matrix. Depending on the fabrication process of the filled elastomer of interest, however, this assumption may not be appropriate. As elaborated below in Sect. 7, this assumption is indeed appropriate for the prototypical case when the liquid inclusions are initially *spherical* in shape.

Constitutive Behavior of the Liquid Inclusions Granted the absence of inertia, the liquid making up the inclusions is presumed to behave as an elastic dielectric fluid. For clarity of presentation, we consider in particular that the electromechanical behavior of the liquid inclusions is characterized by the free energy

$$W_{\mathtt{i}}^{j}(\mathbf{X},\mathbf{F},\mathbf{E}) = r_{\mathtt{i}}^{j}(\mathbf{X})J + \frac{\Lambda_{\mathtt{i}}}{2}(J-1)^2 - \frac{\varepsilon_{\mathtt{i}}}{2}J\mathbf{F}^{-T}\mathbf{E}\cdot\mathbf{F}^{-T}\mathbf{E} \quad j = 1, 2, \ldots, M, \tag{26}$$

where, as will become apparent below in Sect. 7, $r_{\mathtt{i}}^{j}(\mathbf{X})$ shall stand for the pressure—which is *not* necessarily zero due to the possible presence of initial interfacial forces—that the liquid within the jth inclusion is subjected to in the initial configuration, when $\mathbf{F} = \mathbf{I}$ and $\mathbf{E} = \mathbf{0}$, while $\Lambda_{\mathtt{i}} \geq 0$ and $\varepsilon_{\mathtt{i}} \geq \varepsilon_0$ denote the initial first Lamé constant (or bulk modulus since $\mu_{\mathtt{i}} = 0$) and the initial permittivity of the liquid, respectively.

Remark 1.9 Given the constitutive prescription (26), the case of an incompressible liquid corresponds to setting $\Lambda_{\mathtt{i}} = +\infty$, while the case of a conducting liquid corresponds to setting $\varepsilon_{\mathtt{i}} = +\infty$.

Remark 1.10 All M inclusions are assumed to be made of the same liquid, thus the unique values of $\Lambda_{\mathtt{i}}$ and $\varepsilon_{\mathtt{i}}$ in (26). However, because each inclusion is allowed to have its own initial geometry, and thus its own initial size, the term $r_{\mathtt{i}}^{j}(\mathbf{X})$ in (26) describing the residual stress within the inclusions may be different for each inclusion.

Pointwise Constitutive Behavior of the Bulk Given the indicator functions (1) for the inclusions and the free energies (25) and (26) for the matrix and the inclusions, the pointwise free energy for the bulk of the body can be compactly written as

$$\begin{aligned} W(\mathbf{X},\mathbf{F},\mathbf{E}) =& r_{\mathtt{i}}(\mathbf{X})J + \\ & \frac{\mu(\mathbf{X})}{2}[\mathbf{F}\cdot\mathbf{F}-3] - \mu(\mathbf{X})\ln J + \frac{\Lambda(\mathbf{X})}{2}(J-1)^2 - \\ & \frac{\varepsilon(\mathbf{X})}{2}J\mathbf{F}^{-T}\mathbf{E}\cdot\mathbf{F}^{-T}\mathbf{E}, \end{aligned} \tag{27}$$

where

$$\begin{cases} r_{\mathtt{i}}(\mathbf{X}) = \sum_{j=1}^{M} \theta_0^{\mathtt{i},j}(\mathbf{X}) r_{\mathtt{i}}^{j}(\mathbf{X}) \\ \mu(\mathbf{X}) = \left(1 - \theta_0^{\mathtt{i}}(\mathbf{X})\right) \mu_{\mathtt{m}} \\ \Lambda(\mathbf{X}) = \left(1 - \theta_0^{\mathtt{i}}(\mathbf{X})\right) \Lambda_{\mathtt{m}} + \theta_0^{\mathtt{i}}(\mathbf{X}) \Lambda_{\mathtt{i}} \\ \varepsilon(\mathbf{X}) = \left(1 - \theta_0^{\mathtt{i}}(\mathbf{X})\right) \varepsilon_{\mathtt{m}} + \theta_0^{\mathtt{i}}(\mathbf{X}) \varepsilon_{\mathtt{i}} \end{cases}.$$

It then follows that the total first Piola-Kirchhoff stress tensor $\mathbf{S}$ and the Lagrangian electric displacement $\mathbf{D}$ at any material point in the bulk are given by the relations

$$\begin{aligned} \mathbf{S}(\mathbf{X}) = \frac{\partial W}{\partial \mathbf{F}}(\mathbf{X}, \mathbf{F}, \mathbf{E}) = & r_{\mathtt{i}}(\mathbf{X}) J \mathbf{F}^{-T} + \\ & \mu(\mathbf{X}) \left(\mathbf{F} - \mathbf{F}^{-T}\right) + \Lambda(\mathbf{X})(J-1) J \mathbf{F}^{-T} + \\ & \varepsilon(\mathbf{X}) J \mathbf{F}^{-T} \mathbf{E} \otimes \mathbf{F}^{-1} \mathbf{F}^{-T} \mathbf{E} - \\ & \frac{\varepsilon(\mathbf{X})}{2} (\mathbf{F}^{-T} \mathbf{E} \cdot \mathbf{F}^{-T} \mathbf{E}) J \mathbf{F}^{-T}, \quad \mathbf{X} \in \Omega_0 \setminus \Gamma_0 \end{aligned} \tag{28}$$

and

$$\mathbf{D}(\mathbf{X}) = -\frac{\partial W}{\partial \mathbf{E}}(\mathbf{X}, \mathbf{F}, \mathbf{E}) = \varepsilon(\mathbf{X}) J \mathbf{F}^{-1} \mathbf{F}^{-T} \mathbf{E}, \quad \mathbf{X} \in \Omega_0 \setminus \Gamma_0. \tag{29}$$

Remark 1.11 In the limit of small deformations and moderate electric fields[2]—when $\mathbf{H} = \mathbf{F} - \mathbf{I}$ is of $O(\zeta)$, $\mathbf{E}$ is of $O(\zeta^{1/2})$, and $\zeta \searrow 0$—the coupled constitutive response (28) and (29) reduces asymptotically to

$$\begin{aligned} \mathbf{S}(\mathbf{X}) = & r_{\mathtt{i}}(\mathbf{X}) \mathbf{I} - r_{\mathtt{i}}(\mathbf{X}) \mathbf{H}^T + r_{\mathtt{i}}(\mathbf{X})(\operatorname{tr} \mathbf{H}) \mathbf{I} + \\ & \mu(\mathbf{X}) \left(\mathbf{H} + \mathbf{H}^T\right) + \Lambda(\mathbf{X})(\operatorname{tr} \mathbf{H}) \mathbf{I} + \\ & \varepsilon(\mathbf{X}) \mathbf{E} \otimes \mathbf{E} - \frac{\varepsilon(\mathbf{X})}{2} (\mathbf{E} \cdot \mathbf{E}) \mathbf{I} + O(\zeta^2) \end{aligned} \tag{30}$$

and

$$\mathbf{D}(\mathbf{X}) = \varepsilon(\mathbf{X}) \mathbf{E} + O(\zeta^{3/2}). \tag{31}$$

The corresponding total Cauchy stress tensor $\mathbf{T} = J^{-1} \mathbf{S} \mathbf{F}^T$ and Eulerian electric displacement $\mathbf{d} = J^{-1} \mathbf{F} \mathbf{D}$ are given by

[2] For studies of this fundamental limit, see, e.g., Stratton (1941), Tian et al. (2012), Lefèvre and Lopez-Pamies (2014), Spinelli et al. (2015).

$$\begin{aligned}\mathbf{T}(\mathbf{x}) =& r_{\mathtt{i}}(\mathbf{x})\mathbf{I}+\\ &\mu(\mathbf{x})\left(\mathbf{H}+\mathbf{H}^T\right)+\Lambda(\mathbf{x})(\operatorname{tr}\mathbf{H})\mathbf{I}+\\ &\varepsilon(\mathbf{x})\mathbf{E}\otimes\mathbf{E}-\frac{\varepsilon(\mathbf{x})}{2}(\mathbf{E}\cdot\mathbf{E})\,\mathbf{I}+O(\zeta^2)\end{aligned} \tag{32}$$

and

$$\mathbf{d}(\mathbf{x})=\varepsilon(\mathbf{x})\mathbf{E}+O(\zeta^{3/2}). \tag{33}$$

Three key features are now immediate. First, in the initial configuration, when $\mathbf{x}=\mathbf{X}$, $\mathbf{F}=\mathbf{I}$, and $\mathbf{E}=\mathbf{0}$, the Lagrangian relations (30) and (31) and the Eulerian relations (32) and (33) reduce to

$$\begin{cases}\mathbf{S}(\mathbf{X})=r_{\mathtt{i}}(\mathbf{X})\mathbf{I}\\ \mathbf{D}(\mathbf{X})=\mathbf{0}\end{cases}\quad\text{and}\quad\begin{cases}\mathbf{T}(\mathbf{x})=r_{\mathtt{i}}(\mathbf{x})\mathbf{I}\\ \mathbf{d}(\mathbf{x})=\mathbf{0}\end{cases},$$

which indicate that the inclusions (but *not* the matrix) have a hydrostatic residual stress. They also indicate that there is no residual polarization. Second, the stress (30) is *not* symmetric as it does not depend only on the symmetric part of $\mathbf{H}$, but also on $\mathbf{H}$ itself. Third, the total first Piola-Kirchhoff stress (30) does *not* coincide with the total Cauchy stress (32) to $O(||\mathbf{H}||)$. As discussed at length by Ghosh and Lopez-Pamies (2022) and also as elaborated below, these three non-standard features are direct consequences of the presence of a residual stress, which in turn is a direct consequence of the presence of interfacial forces.

Remark 1.12 Thanks to the objectivity of the free energies (25) and (26), the constitutive relation (28) satisfies automatically the balance of angular momentum $(24)_1$ in the bulk.

6.2 Constitutive Behavior of the Solid/liquid Interfaces

Next, we turn to the constitutive description of the interfaces. Similar to the elastomeric matrix and liquid inclusions, we also consider that under the quasistatic deformations and quasistatic electric fields of interest here any (electric or mechanical) interfacial dissipative phenomena is negligible and hence presume the interfaces to exhibit an elastic dielectric behavior. Specifically, we consider that the interface first Piola-Kirchhoff stress tensor $\widehat{\mathbf{S}}$ and the Lagrangian interface electric displacement $\widehat{\mathbf{D}}$ are given by relations of the form

$$\widehat{\mathbf{S}}(\mathbf{X})=\frac{\partial\widehat{W}}{\partial\widehat{\mathbf{F}}}(\widehat{\mathbf{F}},\widehat{\mathbf{E}}),\quad \mathbf{X}\in\Gamma_0 \tag{34}$$

and

$$\widehat{\mathbf{D}}(\mathbf{X}) = -\frac{\partial \widehat{W}}{\partial \widehat{\mathbf{E}}}(\widehat{\mathbf{F}}, \widehat{\mathbf{E}}), \quad \mathbf{X} \in \Gamma_0 \tag{35}$$

in terms of a suitably well-behaved interface free energy (per unit initial area) $\widehat{W}(\widehat{\mathbf{F}}, \widehat{\mathbf{E}})$, where we recall that $\widehat{\mathbf{F}}$ stands for the interface deformation gradient (2) and

$$\widehat{\mathbf{E}}(\mathbf{X}) = \widehat{\mathbf{I}}\mathbf{E}(\mathbf{X})$$

is the *Lagrangian interface electric field.*

In the sequel, for clarity of presentation, we will restrict attention to the ideal-elastic-dielectric-type interface free energy

$$\widehat{W}(\widehat{\mathbf{F}}, \widehat{\mathbf{E}}) = \widehat{\gamma}\widehat{J} + \frac{\widehat{\mu}}{2}\left[\widehat{\mathbf{F}} \cdot \widehat{\mathbf{F}} - 2\right] - \widehat{\mu} \ln \widehat{J} + \frac{\widehat{\Lambda}}{2}(\widehat{J} - 1)^2 - \frac{\widehat{\varepsilon}}{2}\widehat{J}\widehat{\mathbf{F}}^{-T}\widehat{\mathbf{E}} \cdot \widehat{\mathbf{F}}^{-T}\widehat{\mathbf{E}}. \tag{36}$$

In this constitutive prescription, the material constant $\widehat{\gamma} \geq 0$ describes the initial *surface tension* on the solid/liquid interfaces under consideration. On the other hand, $\widehat{\mu} \geq 0$, $\widehat{\Lambda} \geq 0$, and $\widehat{\varepsilon} \geq 0$ can be viewed as the initial *interface Lamé constants* and the initial *interface permittivity*. All three material constants $\widehat{\gamma}$, $\widehat{\mu}$, $\widehat{\Lambda}$ have units of $force/length$. On the other hand, the material constant $\widehat{\varepsilon}$ has units of $force \times length/voltage^2$.

A direct calculation shows that the interface first Piola-Kirchhoff stress tensor (34) and the Lagrangian interface electric displacement (35) associated with the interface free energy (36) are given by

$$\begin{aligned}\widehat{\mathbf{S}}(\mathbf{X}) = \frac{\partial \widehat{W}}{\partial \widehat{\mathbf{F}}}(\widehat{\mathbf{F}}, \widehat{\mathbf{E}}) =& \widehat{\gamma}\widehat{J}\widehat{\mathbf{F}}^{-T} + \\ & \widehat{\mu}(\widehat{\mathbf{F}} - \widehat{\mathbf{F}}^{-T}) + \widehat{\Lambda}(\widehat{J} - 1)\widehat{J}\widehat{\mathbf{F}}^{-T} + \\ & \widehat{\varepsilon}\widehat{J}\widehat{\mathbf{F}}^{-T}\widehat{\mathbf{E}} \otimes \widehat{\mathbf{F}}^{-1}\widehat{\mathbf{F}}^{-T}\widehat{\mathbf{E}} - \\ & \frac{\widehat{\varepsilon}}{2}(\widehat{\mathbf{F}}^{-T}\widehat{\mathbf{E}} \cdot \widehat{\mathbf{F}}^{-T}\widehat{\mathbf{E}})\widehat{J}\widehat{\mathbf{F}}^{-T}, \quad \mathbf{X} \in \Gamma_0\end{aligned} \tag{37}$$

and

$$\widehat{\mathbf{D}}(\mathbf{X}) = -\frac{\partial \widehat{W}}{\partial \widehat{\mathbf{E}}}(\widehat{\mathbf{F}}, \widehat{\mathbf{E}}) = \widehat{\varepsilon}\widehat{J}\widehat{\mathbf{F}}^{-1}\widehat{\mathbf{F}}^{-T}\widehat{\mathbf{E}}, \quad \mathbf{X} \in \Gamma_0. \tag{38}$$

Remark 1.13 Another direct calculation shows that the interface Cauchy stress tensor $\widehat{\mathbf{T}} = \widehat{J}^{-1}\widehat{\mathbf{S}}\widehat{\mathbf{F}}^T$ associated with the free energy (36) reads

$$\begin{aligned}\widehat{\mathbf{T}}(\mathbf{x}) =& \widehat{\gamma}\widehat{\mathbf{i}} + \widehat{\mu}(\widehat{J}^{-1}\widehat{\mathbf{F}}\widehat{\mathbf{F}}^T - \widehat{\mathbf{i}}) + \widehat{\Lambda}(\widehat{J} - 1)\widehat{\mathbf{i}} + \\ & \widehat{\varepsilon}\widehat{\mathbf{F}}^{-T}\widehat{\mathbf{E}} \otimes \widehat{\mathbf{F}}^{-T}\widehat{\mathbf{E}} - \frac{\widehat{\varepsilon}}{2}(\widehat{\mathbf{F}}^{-T}\widehat{\mathbf{E}} \cdot \widehat{\mathbf{F}}^{-T}\widehat{\mathbf{E}})\widehat{\mathbf{i}}, \quad \mathbf{x} \in \Gamma.\end{aligned} \tag{39}$$

This expression makes it plain that the constitutive relation (37) utilized here to describe the electromechanical behavior of the interfaces generalizes in three counts the basic constitutive relation of constant surface-tension stress

$$\widehat{\mathbf{T}}(\mathbf{x}) = \widehat{\gamma}\widehat{\mathbf{i}}.$$

Specifically, the constitutive relation (39) includes Neo-Hookean-type deviatoric elasticity, via the term $\widehat{\mu}(\widehat{J}^{-1}\widehat{\mathbf{F}}\widehat{\mathbf{F}}^T - \widehat{\mathbf{i}})$, and not just surface tension. It also accounts for a surface tension that is not necessarily a constant but instead one that depends on the deformation of the interface via the term $\widehat{\Lambda}(\widehat{J} - 1)\widehat{\mathbf{i}}$. Finally, the last two terms in the constitutive relation (39) describe the presence of an interfacial polarization.

Remark 1.14 Thanks to the objectivity of the free energy (36), the constitutive relation (37) satisfies automatically the balance of angular momentum $(24)_2$ on the interfaces.

7 Governing Equations

7.1 Boundary Conditions

In terms of the external stimuli applied to the body, we have already described the source terms of bulk and interface charges Q and $\widehat{Q}$ and the mechanical body force $\mathbf{B}$. We now describe the external stimuli applied on the boundary of the body.

From an electric point of view, we take that the body is immersed in a surrounding space, e.g., air, where there is a heterogeneous electric field $\overline{\mathbf{E}}(\mathbf{X})$ and corresponding electric displacement $\overline{\mathbf{D}}(\mathbf{X})$ that result by the use of electrodes, where a surface charge density per unit initial area $\overline{Q}$ is applied, and/or the nearby presence of polarized bodies and by the interaction of these with the body. We then have the boundary condition

$$(\mathbf{I} - \mathbf{N} \otimes \mathbf{N})\mathbf{E} = (\mathbf{I} - \mathbf{N} \otimes \mathbf{N})\overline{\mathbf{E}}, \quad \mathbf{X} \in \partial\Omega_0, \tag{40}$$

or, equivalently,

$$\mathbf{D} \cdot \mathbf{N} = -\overline{Q} + \overline{\mathbf{D}} \cdot \mathbf{N}, \quad \mathbf{X} \in \partial\Omega_0,$$

over the entirety of the boundary of the domain occupied by the body.

From a mechanical point of view, on a portion $\partial\Omega_0^{\mathcal{D}}$ of the boundary $\partial\Omega_0$, the deformation field $\mathbf{y}$ is taken to be given by a known function $\overline{\mathbf{y}}(\mathbf{X})$, while the complementary part of the boundary $\partial\Omega_0^{\mathcal{N}} = \partial\Omega_0 \setminus \partial\Omega_0^{\mathcal{D}}$ is subjected to a prescribed mechanical traction $\overline{\mathbf{t}}_{\mathrm{m}}(\mathbf{X})$. Precisely,

$$\mathbf{y} = \overline{\mathbf{y}}, \quad \mathbf{X} \in \partial\Omega_0^{\mathcal{D}} \qquad \text{and} \qquad \mathbf{S}\mathbf{N} = \overline{\mathbf{t}}_{\mathrm{m}} + \overline{\mathbf{S}}_{\mathrm{e}}\mathbf{N}, \quad \mathbf{X} \in \partial\Omega_0^{\mathcal{N}}. \tag{41}$$

In this last expression, $\overline{\mathbf{S}}_{\mathrm{e}}$ stands for the Maxwell stress outside of the body. In the case when the body is surrounded by air,

$$\overline{\mathbf{S}}_{\mathrm{e}}=\mathbf{F}^{-T}\mathbf{E}\otimes\mathbf{D}-\frac{J\varepsilon_0}{2}\left(\mathbf{F}^{-T}\mathbf{E}\cdot\mathbf{F}^{-T}\mathbf{E}\right)\mathbf{F}^{-T},$$

where $\mathbf{D}=\varepsilon_0 J\mathbf{F}^{-1}\mathbf{F}^{-T}\mathbf{E}$ and where we remark that the deformation gradient $\mathbf{F}$ in the air refers to any suitably well-behaved extension of the deformation gradient $\mathbf{F}$ in the body.

7.2 *The Choice of Independent Fields*

At this stage, all that remains to formulate a mathematically closed system of governing equations is to identify the independent fields that we wish to solve for. Arguably, the most expedient choice of independent fields for the problem at hand is the deformation field

$$\mathbf{y}(\mathbf{X})$$

and the electric potential

$$\Phi(\mathbf{X}).$$

Recall that the rest of fields can be written in terms of these two as follows:

$$\begin{cases}\mathbf{F}=\nabla\mathbf{y}\\ \widehat{\mathbf{F}}=\mathbf{F}\widehat{\mathbf{I}}\\ \mathbf{E}=\nabla\Phi\\ \widehat{\mathbf{E}}=\widehat{\mathbf{I}}\mathbf{E}\end{cases},\quad \begin{cases}\mathbf{S}=\dfrac{\partial W}{\partial\mathbf{F}}(\mathbf{X},\mathbf{F},\mathbf{E})\\ \widehat{\mathbf{S}}=\dfrac{\partial\widehat{W}}{\partial\widehat{\mathbf{F}}}(\widehat{\mathbf{F}},\widehat{\mathbf{E}})\\ \mathbf{D}=-\dfrac{\partial W}{\partial\mathbf{E}}(\mathbf{X},\mathbf{F},\mathbf{E})\\ \widehat{\mathbf{D}}=-\dfrac{\partial\widehat{W}}{\partial\widehat{\mathbf{E}}}(\widehat{\mathbf{F}},\widehat{\mathbf{E}})\end{cases},\quad \begin{cases}\rho=J^{-1}\rho_0\\ \mathbf{T}=J^{-1}\mathbf{S}\mathbf{F}^T\\ \widehat{\mathbf{T}}=\widehat{J}^{-1}\widehat{\mathbf{S}}\widehat{\mathbf{F}}^T\\ \mathbf{d}=J^{-1}\mathbf{F}\mathbf{D}\\ \widehat{\mathbf{d}}=\widehat{J}^{-1}\widehat{\mathbf{F}}\widehat{\mathbf{D}}\\ \mathbf{e}=\mathbf{F}^{-T}\mathbf{E}\\ \widehat{\mathbf{e}}=\widehat{\mathbf{F}}^{-T}\widehat{\mathbf{E}}\\ \phi=\Phi\end{cases}$$

7.3 *The Strong Form of the Governing Equations*

Granted the choice of the deformation field $\mathbf{y}(\mathbf{X})$ and the electric potential $\Phi(\mathbf{X})$ as the independent fields, the equation of conservation of mass (5) and Faraday's law (14) are automatically satisfied. Granted the use of objective free energies $W(\mathbf{X},\mathbf{F},\mathbf{E})$ and objective interface free energies $\widehat{W}(\widehat{\mathbf{F}},\widehat{\mathbf{E}})$, such as (27) and (36), the equations of balance of angular momentum (24) are also automatically satisfied. Thus, Gauss's law (11) and the balance of linear momentum (21) are the only two sets of balance principles that need to be solved.

Substitution of the constitutive relations (28), (29), (37) and (38) for the bulk and interfaces in Gauss's law (11), the balance of linear momentum (21), and the boundary conditions (40) and (41), together with use of the notation $\overline{\mathbf{E}} = \nabla\overline{\Phi}(\mathbf{X})$, yields the following governing equations

$$\begin{cases} \mathrm{Div}\left[-\frac{\partial W}{\partial \mathbf{E}}(\mathbf{X}, \nabla\mathbf{y}, \nabla\Phi)\right] = Q, \quad \mathbf{X} \in \Omega_0 \setminus \Gamma_0 \\ \widehat{\mathrm{Div}}\left[-\frac{\partial \widehat{W}}{\partial \widehat{\mathbf{E}}}(\widehat{\nabla}\mathbf{y}, \widehat{\nabla}\Phi)\right] - [\![-\frac{\partial W}{\partial \mathbf{E}}(\mathbf{X}, \nabla\mathbf{y}, \nabla\Phi)]\!]\widehat{\mathbf{N}} = \widehat{Q}, \quad \mathbf{X} \in \Gamma_0 \\ \Phi(\mathbf{X}) = \overline{\Phi}(\mathbf{X}), \quad \mathbf{X} \in \partial\Omega_0 \end{cases} \tag{42}$$

and

$$\begin{cases} \mathrm{Div}\left[\frac{\partial W}{\partial \mathbf{F}}(\mathbf{X}, \nabla\mathbf{y}, \nabla\Phi)\right] + \mathbf{B} = \mathbf{0}, \quad \mathbf{X} \in \Omega_0 \setminus \Gamma_0 \\ \widehat{\mathrm{Div}}\left[\frac{\partial \widehat{W}}{\partial \widehat{\mathbf{F}}}(\widehat{\nabla}\mathbf{y}, \widehat{\nabla}\Phi)\right] - [\![\frac{\partial W}{\partial \mathbf{F}}(\mathbf{X}, \nabla\mathbf{y}, \nabla\Phi)]\!]\widehat{\mathbf{N}} = \mathbf{0}, \quad \mathbf{X} \in \Gamma_0 \\ \mathbf{y}(\mathbf{X}) = \overline{\mathbf{y}}(\mathbf{X}), \quad \mathbf{X} \in \partial\Omega_0^{\mathcal{D}} \\ \left[\frac{\partial W}{\partial \mathbf{F}}(\mathbf{X}, \nabla\mathbf{y}, \nabla\Phi)\right]\mathbf{N} = \overline{\mathbf{t}}_{\mathrm{m}}(\mathbf{X}) + \overline{\mathbf{S}}_{\mathrm{e}}\mathbf{N}, \quad \mathbf{X} \in \partial\Omega_0^{\mathcal{N}} \end{cases} \tag{43}$$

for the deformation field $\mathbf{y}(\mathbf{X})$ and the electric potential $\Phi(\mathbf{X})$.

Equations (42) and (43) constitute a generalization of the classical governing equations for heterogenous elastic dielectrics under quasistatic deformations and quasistatic electric fields that accounts for: (i) the presence of residual stresses (in the inclusions) and (ii) jump conditions across material (solid/liquid) interfaces that are not algebraic but, instead, are described by PDEs (partial differential equations) that result from the presence of interfacial polarization and forces.

7.4 Residual Stresses

In the initial configuration, prior to the application of the charges Q, $\widehat{Q}$, the body force $\mathbf{B}$, and the boundary conditions $\overline{\mathbf{y}}$, $\overline{\mathbf{t}}_{\mathrm{m}}$, and $\overline{\mathbf{S}}_{\mathrm{e}}$, the deformation field $\mathbf{y}(\mathbf{X}) = \mathbf{X}$ and the electric potential $\Phi(\mathbf{X}) = 0$ and hence the governing Eqs. (42) and (43) reduce to

$$\begin{cases} \mathrm{Div}\,[r_{\mathrm{i}}(\mathbf{X})\mathbf{I}] = \mathbf{0}, \quad \mathbf{X} \in \Omega_0 \setminus \Gamma_0 \\ \widehat{\gamma}\,\widehat{\mathrm{Div}}\,\widehat{\mathbf{I}} - [\![r_{\mathrm{i}}(\mathbf{X})]\!]\widehat{\mathbf{N}} = \mathbf{0}, \quad \mathbf{X} \in \Gamma_0 \end{cases}, \tag{44}$$

which can be viewed as the definition of the hydrostatic residual stress $r_{\mathrm{i}}(\mathbf{X})$ within the inclusions required to balance out the interfacial forces. Recognizing that $[\![r_{\mathrm{i}}(\mathbf{X})]\!] = r_{\mathrm{i}}(\mathbf{X})$ and that

$$\widehat{\mathrm{Div}}\,\widehat{\mathbf{I}} = -\nabla(\widehat{\mathbf{N}} \otimes \widehat{\mathbf{N}}) \cdot \widehat{\mathbf{I}} = -(\widehat{\mathbf{I}} \cdot \nabla\widehat{\mathbf{N}})\widehat{\mathbf{N}} = -(\mathrm{tr}\,\widehat{\nabla}\widehat{\mathbf{N}})\widehat{\mathbf{N}} = 2\kappa\widehat{\mathbf{N}}$$

in terms of the mean curvature $\kappa = -\text{tr}\,\widehat{\nabla}\widehat{\mathbf{N}}/2$ of the interfaces, Eq. (44) can be rewritten more explicitly as

$$\begin{cases} \nabla r_{\mathtt{i}}(\mathbf{X}) = \mathbf{0}, & \mathbf{X} \in \Omega_0 \setminus \Gamma_0 \\ r_{\mathtt{i}}(\mathbf{X}) = 2\kappa\widehat{\gamma}, & \mathbf{X} \in \Gamma_0 \end{cases}. \tag{45}$$

The PDE $(45)_1$ states that the hydrostatic residual stress $r_{\mathtt{i}}(\mathbf{X})$ must be *constant*—possibly a different constant—within each inclusion. In view of the boundary condition $(45)_2$, which is nothing more than the standard Young-Laplace equation, a solution to the boundary-value problem (45) then only exists for the case when all M inclusions have shapes of *constant mean curvature* (Kenmotsu (2003)), for only then $(45)_2$ is consistent with $(45)_1$. Physically, as alluded to in Remark 1.8, this result implies that to deal with liquid inclusions of general initial shape, one would have to account for residual stresses in the elastomeric matrix and not just within the inclusions.

Remark 1.15 The prototypical case of elastomers filled with liquid inclusions that have constant mean curvature—and hence for which the governing Eqs. (42) and (43) apply—is that of elastomers that are filled with liquid inclusions that are initially *spherical* in shape. For these, the solution to (45) simply reads

$$r_{\mathtt{i}}(\mathbf{X}) = -\sum_{j=1}^{M} \theta_0^{\mathtt{i},j}(\mathbf{X})\frac{2\widehat{\gamma}}{A_j},$$

where A_j denotes the initial radius of the jth inclusion.

Acknowledgements Support for this work by the National Science Foundation through the Grant DMREF–1922371 is gratefully acknowledged.

Appendix A. Gauss's law in Lagrangian form

On substitution of the definitions $Q = Jq$, $\mathbf{D} = J\mathbf{F}^{-1}\mathbf{d}$, $\widehat{Q} = \widehat{J}\widehat{q}$, and $\widehat{\mathbf{D}} = \widehat{J}\widehat{\mathbf{F}}^{-1}\widehat{\mathbf{d}}$ in the Eulerian form (10) of Gauss's law, we have

$$\begin{cases} \dfrac{\partial}{\partial x_k}\left[J^{-1}F_{km}D_m\right] = J^{-1}Q, & \mathbf{x} \in \Omega \setminus \Gamma \\ \dfrac{\partial}{\partial x_l}\left[\widehat{J}^{-1}\widehat{F}_{km}\widehat{D}_m\right]\widehat{i}_{kl} - [[J^{-1}F_{km}D_m]]\widehat{n}_k = \widehat{J}^{-1}\widehat{Q}, & \mathbf{x} \in \Gamma \\ \dfrac{\partial}{\partial x_k}\left[J^{-1}F_{km}D_m\right] = 0, & \mathbf{x} \in \mathbb{R}^3 \setminus \Omega \\ [[J^{-1}F_{km}D_m]]\, n_k = 0, & \mathbf{x} \in \partial\Omega \end{cases}, \tag{46}$$

where, again, by $\mathbf{F}$ in $\mathbb{R}^3 \setminus \Omega$ we mean any suitably well-behaved extension to $\mathbb{R}^3$ of the deformation gradient $\mathbf{F}$ in the body. Given that $\mathrm{div}\left[J^{-1}\mathbf{F}^T\right] = \widehat{\mathrm{div}}\left[\widehat{J}^{-1}\widehat{\mathbf{F}}^T\right] = \mathbf{0}$, Eq. (46) simplify to

$$
\begin{cases}
J^{-1}F_{km}\dfrac{\partial D_m}{\partial x_k} = J^{-1}Q, & \mathbf{x} \in \Omega \setminus \Gamma \\
\widehat{J}^{-1}\widehat{F}_{km}\dfrac{\partial \widehat{D}_m}{\partial x_l}\widehat{i}_{kl} - [\![J^{-1}F_{km}D_m]\!]\,\widehat{n}_k = \widehat{J}^{-1}\widehat{Q}, & \mathbf{x} \in \Gamma \\
J^{-1}F_{km}\dfrac{\partial D_m}{\partial x_k} = 0, & \mathbf{x} \in \mathbb{R}^3 \setminus \Omega \\
[\![J^{-1}F_{km}D_m]\!]\,n_k = 0, & \mathbf{x} \in \partial\Omega
\end{cases}.
$$

By employing now the chain rule and the identities $\mathbf{n} = |J\mathbf{F}^{-T}\mathbf{N}|^{-1}\, J\mathbf{F}^{-T}\mathbf{N}$, $\widehat{\mathbf{F}}^{-1} = \mathbf{F}^{-1}\widehat{\mathbf{i}}$, and $\widehat{\mathbf{n}} = \widehat{J}^{-1}\, J\mathbf{F}^{-T}\widehat{\mathbf{N}}$ together with the fact that $J^{\mathrm{i}}\mathbf{F}^{\mathrm{i}\,-T}\widehat{\mathbf{N}} = J^{\mathrm{m}}\mathbf{F}^{\mathrm{m}\,-T}\widehat{\mathbf{N}}$, we obtain

$$
\begin{cases}
\dfrac{\partial D_m}{\partial X_n}F^{-1}_{nk}F_{km} = Q, & \mathbf{X} \in \Omega_0 \setminus \Gamma_0 \\
\dfrac{\partial \widehat{D}_m}{\partial X_n}\widehat{F}^{-1}_{nk}\widehat{F}_{km} - [\![D_k]\!]\,\widehat{N}_k = \widehat{Q}, & \mathbf{X} \in \Gamma_0 \\
\dfrac{\partial D_m}{\partial X_n}F^{-1}_{nk}F_{km} = 0, & \mathbf{X} \in \mathbb{R}^3 \setminus \Omega_0 \\
[\![D_k]\!]\,\widehat{N}_k = 0, & \mathbf{X} \in \partial\Omega_0
\end{cases}
$$

Finally, recognizing that $\widehat{\mathbf{F}}^{-1}\widehat{\mathbf{F}} = \widehat{\mathbf{I}}$, Gauss's law in Lagrangian form (11) readily follows:

$$
\begin{cases}
\dfrac{\partial D_m}{\partial X_m} = Q, & \mathbf{X} \in \Omega_0 \setminus \Gamma_0 \\
\dfrac{\partial \widehat{D}_m}{\partial X_n}\widehat{I}_{mn} - [\![D_k]\!]\,\widehat{N}_k = \widehat{Q}, & \mathbf{X} \in \Gamma_0 \\
\dfrac{\partial D_m}{\partial X_m} = 0, & \mathbf{X} \in \mathbb{R}^3 \setminus \Omega_0 \\
[\![D_k]\!]\,\widehat{N}_k = 0, & \mathbf{X} \in \partial\Omega_0
\end{cases}
$$

Appendix B. Faraday's law in Lagrangian form

Direct use of the definition $\mathbf{E} = \mathbf{F}^T\mathbf{e}$ and the transformation rule $\mathrm{d}\mathbf{x} = \mathbf{F}\mathrm{d}\mathbf{X}$ for material line elements allows to recast the integral form (12) of Faraday's law as

$$\int_{\partial\Sigma} \mathbf{e}\cdot \mathrm{d}\mathbf{x} = \int_{\partial\Sigma_0} \mathbf{E}\cdot \mathrm{d}\mathbf{X} = 0. \tag{47}$$

By making use of Stokes's theorem

$$\int_{\Sigma_0} (\mathrm{Curl}\,\mathbf{E})\cdot \widetilde{\mathbf{N}}\,\mathrm{d}\mathbf{X} = \int_{\partial\Sigma_0} \mathbf{E}\cdot \mathrm{d}\mathbf{X} + \int_{\partial\mathcal{S}_0} [\![\mathbf{E}]\!]\cdot \mathrm{d}\mathbf{X},$$

this time around in the initial configuration, the Lagrangian localized form (14) of Faraday's law readily follows from (47).

References

Bartlett, M. D., Kazem, N., Powell-Palm, M. J., Huang, X., Sun, W., Malen, J. A., & Majidi, C. (2017). High thermal conductivity in soft elastomers with elongated liquid metal inclusions. *Proceedings of the National Academy of Sciences, 114*, 2143–2148.

Díaz, J. C., Francfort, G. A., Lopez-Pamies, O., & Mora, M. G. (2023) Liquid filled elastomers: From linearization to elastic enhancement.

do Carmo, M. P. (2016) *Differential geometry of curves and surfaces*. Dover

Dorfmann, A., & Ogden, R. W. (2005). Nonlinear electroelasticity. *Acta Mechanica, 174*, 167–183.

Ghosh, K., & Lopez-Pamies, O. (2022). Elastomers filled with liquid inclusions: Theory, numerical implementation, and some basic results. *Journal of the Mechanics and Physics of Solids, 166*, 104930.

Ghosh, K., Lefèvre, V., & Lopez-Pamies, O. (2023a) Homogenization of elastomers filled with liquid inclusions: The small-deformation limit. *Journal of Elasticity*.

Ghosh, K., Lefèvre, V., & Lopez-Pamies, O. (2023b). The effective shear modulus of a random isotropic suspension of monodisperse liquid n-spheres: From the dilute limit to the percolation threshold. *Soft Matter, 19*, 208–224.

Gibbs, J. W. (1878). On the equilibrium of heterogeneous substances. *Transactions of the Connecticut Academy of Arts and Sciences, 3*, 343–524.

Gurtin, M. E., & Murdoch, A. I. (1975a). A continuum theory of elastic material surfaces. *Archive for Rational Mechanics and Analysis, 57*, 291–323.

Gurtin, M. E., & Murdoch, A. I. (1975b). Addenda to our paper a continuum theory of elastic material surfaces. *Archive for Rational Mechanics and Analysis, 59*, 1–2.

Gurtin, M. E., Weissmüller, J., & Larché, F. (1998). A general theory of curved deformable interfaces in solids at equilibrium. *Philosophical Magazine A, 78*, 1093–1109.

Javili, A., McBride, A., & Steinmann, P. (2013) Thermomechanics of solids with lower-dimensional energetics: On the importance of surface, interface, and curve structures at the nanoscale. A unifying review. *Applied Mechanics Reviews, 65*, 010802.

Kenmotsu, K. (2003). *Surfaces with constant mean curvature*. Providence: American Mathematical Society.

Krichen, S., Liu, L., & Sharma, P. (2019). Liquid inclusions in soft materials: Capillary effect, mechanical stiffening and enhanced electromechanical response. *Journal of the Mechanics and Physics of Solids, 127*, 332–357.

Laplace, P. S. (1806) *Traité de mécanique céleste, Volume 4, Supplémeent au dixième livre*, pp. 1–79.

Lefèvre, V., & Lopez-Pamies, O. (2014). The overall elastic dielectric properties of a suspension of spherical particles in rubber: An exact explicit solution in the small-deformation limit. *Journal of Applied Physics, 116*, 134106.

Lefèvre, V., & Lopez-Pamies, O. (2017a). Nonlinear electroelastic deformations of dielectric elastomer composites: I—Ideal elastic dielectrics. *Journal of the Mechanics and Physics of Solids, 99*, 409–437.
Lefèvre, V., & Lopez-Pamies, O. (2017b). Nonlinear electroelastic deformations of dielectric elastomer composites: II–Non-Gaussian elastic dielectrics. *Journal of the Mechanics and Physics of Solids, 99*, 438–470.
Lefèvre, V., Danas, K., & Lopez-Pamies, O. (2017). A general result for the magnetoelastic response of isotropic suspensions of iron and ferrofluid particles in rubber, with applications to spherical and cylindrical specimens. *Journal of the Mechanics and Physics of Solids, 107*, 343–364.
Lopez-Pamies, O. (2014). Elastic dielectric composites: Theory and application to particle-filled ideal dielectrics. *Journal of the Mechanics and Physics of Solids, 64*, 61–82.
Ogden, R. W. (1997). *Non-linear elastic deformations*. Dover.
Pao, Y. H. (1978). Electromagnetic forces in deformable continua. *Mechanics Today, 4*, 209–306.
Popinet, S. (2018). Numerical models of surface tension. *Annual Review of Fluid Mechanics, 50*, 49–75.
Spinelli, S. A., Lefèvre, V., & Lopez-Pamies, O. (2015). Dielectric elastomer composites: A general closed-form solution in the small-deformation limit. *Journal of the Mechanics and Physics of Solids, 83*, 263–284.
Stratton, J. S. (1941). *Electromagnetic theory*. McGraw-Hill.
Style, R. W., Boltyanskiy, R., Benjamin, A., Jensen, K. E., Foote, H. P., Wettlaufer, J. S., & Dufresne, E. R. (2015a). Stiffening solids with liquid inclusions. *Nature Physics, 11*, 82–87.
Style, R. W., Wettlaufer, J. S., & Dufresne, E. R. (2015b). Surface tension and the mechanics of liquid inclusions in compliant solids. *Soft Matter, 11*, 672–679.
Tian, L., Tevet-Deree, L., deBotton, G., & Bhattacharya, K. (2012) Dielectric elastomer composites. *Journal of the Mechanics and Physics of Solids, 60*, 181–198.
Wang, Y., & Henann, D. L. (2016). Finite-element modeling of soft solids with liquid inclusions. *Extreme Mechanics Letters, 9*, 147–157.
Weatherburn, C. E. (2016). *Differential geometry of three dimensions*. Cambridge University Press.
Young, T. (1805). Iii an essay on the cohesion of fluids. *Philosophical Transactions of the Royal Society, 95*, 9565–9587.
Yun, G., Tang, S. Y., Sun, S., Yuan, D., Zhao, Q., Deng, L., Yan, S., Du, H., Dickey, M. D., & Li, W. (2019). Liquid metal-filled magnetorheological elastomer with positive piezoconductivity. *Nature Communications, 10*, 1300.

Modelling of Homogeneous and Composite Non-linear Electro-Elastic Elastomers

Massimiliano Gei

Abstract We present in a concise way the main features of non-linear electro-elastic modelling of homogeneous and composite elastomers. We focus on some significant examples: three types of actuation of thin dielectric elastomers, search for bifurcations of prestretched membranes, introduction to modelling of two-phase laminates and optimization of performance of a dielectric energy generator that is assumed as a prototypical example of a soft dielectric device.

1 Introduction

Soft electro-elastic devices are commonly composed of thin elastomer films actuated by an electric field generated by a jump in the electric potential across the thickness (Pelrine et al., 1998, 2000). They have found application in many fields of engineering such as biomedical (Calabrese et al., 2018; Carpi et al., 2014; Lu et al., 2016), robotic (Calabrese et al., 2019; Chen et al., 2019; Gupta et al., 2019), mechanical (Conn & Rossiter, 2012; McGough et al., 2014; Wingert et al., 2006), and energy engineering (Bortot et al., 2014; Kornbluh et al., 2011; McKay et al., 2011; Moretti et al., 2019). The goal of this chapter is to introduce in a concise manner the main methods to model different types of electro-elastic devices; in particular, we address:

- three types of actuation of a homogeneous thin dielectric membrane;
- how to tackle bifurcations along an electro-elastic equilibrium path;
- modelling of electro-elastic laminates (a laminate is a significant example of a composite material);

Work done under the auspices of GNFM (Gruppo Nazionale per la Fisica Matematica) of the Italian institute INDAM (Istituto Nazionale di Alta Matematica Francesco Severi).

M. Gei (✉)
Department of Engineering and Architecture, University of Trieste, Trieste, Italy
e-mail: massimiliano.gei@dia.units.it

K. Danas and O. Lopez-Pamies (eds.), *Electro- and Magneto-Mechanics of Soft Solids*,
CISM International Centre for Mechanical Sciences 610,
https://doi.org/10.1007/978-3-031-48351-6_2

- modelling of a prototype of an electro-elastic mechanical-to-electrical energy converter that exploits change in capacitance of a soft dielectric membrane induced by a mechanical action powered by the environment.

For each problem, a list of references are added as a guide for the reader interested to deepen the topic. Overall, the review paper (Lu et al., 2020) and the book (Carpi et al., 2016) are suggested readings for a general overview on modelling of electro-elastic polymers and how these materials are profitably employed in soft actuators and other types of devices.

2 Modelling of Electro-Elastic Materials

We assume that in the stress-free configuration the electro-elastic material occupies a region $B^0 \in \mathbb{R}^3$. After a given deformation χ, the deformed body covers a space $B \in \mathbb{R}^3$. We can define a material particle in B^0 by its position vector $\boldsymbol{x}^0$ that is transformed to $\boldsymbol{x} = \chi(\boldsymbol{x}^0)$ in B; $\boldsymbol{F} = \partial\chi/\partial\boldsymbol{x}^0$ denotes the deformation gradient. The space surrounding the body will be denoted by B^{sur} with the particular case of vacuum being indicated with B^*.

In the deformed configuration B (Fig. 1), we assume that the total stress is denoted by $\boldsymbol{\tau}$ while electric and displacement fields are indicated with $\boldsymbol{E}$ and $\boldsymbol{D}$, respectively. Under the assumption of the absence of both volume free charges and body forces, the governing equations are the following (Dorfmann & Ogden, 2005; McMeeking & Landis, 2005; Suo et al., 2008):

$$\mathrm{div}\boldsymbol{\tau} = \boldsymbol{0}, \quad \boldsymbol{\tau}^T = \boldsymbol{\tau}, \quad \mathrm{div}\boldsymbol{D} = 0, \quad \mathrm{curl}\boldsymbol{E} = \boldsymbol{0}. \tag{1}$$

The last equation implies that $\boldsymbol{E}$ can be derived from the electric potential $\phi(\boldsymbol{x})$ such that $\boldsymbol{E} = -\mathrm{grad}\phi(\boldsymbol{x})$. In vacuum, Eqs. (1) specialise to

$$\mathrm{div}\,\boldsymbol{T}^* = \boldsymbol{0}, \quad \boldsymbol{T}^* = (\boldsymbol{T}^*)^T, \quad \mathrm{div}\boldsymbol{D}^* = 0, \quad \mathrm{curl}\boldsymbol{E}^* = \boldsymbol{0}, \tag{2}$$

where $\boldsymbol{T}^*$ is the Maxwell stress, i.e.

$$\boldsymbol{T}^* = \varepsilon_0\Big(\boldsymbol{E}^* \otimes \boldsymbol{E}^* - \frac{1}{2}(\boldsymbol{E}^* \cdot \boldsymbol{E}^*)\boldsymbol{I}\Big). \tag{3}$$

In Eq. (3), $\varepsilon_0 = 8.85 \cdot 10^{-12}\,\mathrm{F/m}$ is the permittivity of the vacuum and $\boldsymbol{I}$ is the appropriate identity tensor. The electric quantities follow the relationship $\boldsymbol{D}^* = \varepsilon_0\boldsymbol{E}^*$, while Eq. $(2)_4$ implies $\boldsymbol{E}^* = -\mathrm{grad}\phi^*(\boldsymbol{x})$.

In the deformed configuration, we can define jump conditions at the boundaries across a surface discontinuity,

$$\boldsymbol{n} \cdot [[\boldsymbol{D}]] = -\omega, \quad \boldsymbol{n} \times [[\boldsymbol{E}]] = \boldsymbol{0}, \quad [[\boldsymbol{\tau}]]\boldsymbol{n} = \boldsymbol{t}, \tag{4}$$

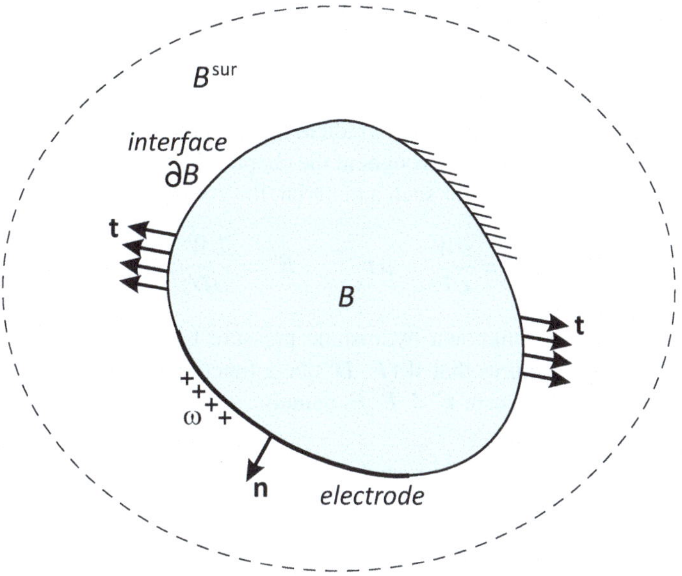

Fig. 1 Sketch of an electro-elastic body B surrounded by vacuum B^{sur} under electro-mechanical loadings

where $[[f]] = f_B - f_{B^{\mathrm{sur}}}$. Here, $\boldsymbol{n}$ is the outward unit normal to B, ω is the surface charge density and $\boldsymbol{t}$ is the mechanical traction. When the discontinuity occurs at the boundary between the elastomer and an adjacent vacuum the above equations specialise as

$$\boldsymbol{n} \cdot \boldsymbol{D} = -\omega + \varepsilon_0 \boldsymbol{E}^* \cdot \boldsymbol{n}, \quad \boldsymbol{n} \times (\boldsymbol{E} - \boldsymbol{E}^*) = \boldsymbol{0}, \quad \boldsymbol{\tau n} = \boldsymbol{t} + \boldsymbol{T}^* \boldsymbol{n}. \tag{5}$$

It could be useful to define nominal or Lagrangian measures. Following the usual argument, the total first Piola-Kirchhoff stress tensor is defined as the two-point tensor

$$\boldsymbol{S} = J \boldsymbol{\tau} \boldsymbol{F}^{-T}, \tag{6}$$

where $J = \det \boldsymbol{F}$, while the Lagrangian electric displacement and electric fields are given by

$$\boldsymbol{D}^0 = J \boldsymbol{F}^{-1} \boldsymbol{D}, \quad \boldsymbol{E}^0 = \boldsymbol{F}^T \boldsymbol{E}, \tag{7}$$

respectively. The governing equations. (1) updated to the reference configuration thus become

$$\mathrm{Div} \boldsymbol{S} = \boldsymbol{0}, \quad \boldsymbol{S F}^T = \boldsymbol{F S}^T, \quad \mathrm{Div} \boldsymbol{D}^0 = 0, \quad \mathrm{Curl} \boldsymbol{E}^0 = \boldsymbol{0}, \tag{8}$$

where differentiation is taken now with respect to coordinates in B^0.

2.1 Electro-Elastic Constitutive Equations

The material considered can be defined by an electro-elastic free-energy function W which is based on the deformation gradient and nominal electrical displacement $\boldsymbol{D}^0$ such that $W = W(\boldsymbol{F}, \boldsymbol{D}^0)$. Throughout the chapter the material is assumed to be incompressible, then $J = 1$. For such a material, the constitutive equations take the form

$$\boldsymbol{S} = \frac{\partial W}{\partial \boldsymbol{F}} - p\boldsymbol{F}^{-T}, \quad \boldsymbol{E}^0 = \frac{\partial W}{\partial \boldsymbol{D}^0}, \tag{9}$$

where p represents an unknown hydrostatic pressure to be defined by boundary conditions. Isotropy requires that $W(\boldsymbol{F}, \boldsymbol{D}^0)$ be a function of both the invariants of the right Cauchy-Green tensor $\boldsymbol{C} = \boldsymbol{F}^T\boldsymbol{F}$, namely

$$I_1 = \mathrm{tr}\boldsymbol{C}, \quad I_2 = \frac{1}{2}\left[(\mathrm{tr}\boldsymbol{C})^2 - \mathrm{tr}(\boldsymbol{C}^2)\right], \quad I_3 = \det\boldsymbol{C} = 1, \tag{10}$$

and those involving both $\boldsymbol{C}$ and $\boldsymbol{D}^0$, i.e.

$$I_4 = \boldsymbol{D}^0 \cdot \boldsymbol{D}^0, \quad I_5 = \boldsymbol{D}^0 \cdot \boldsymbol{C}\boldsymbol{D}^0, \quad I_6 = \boldsymbol{D}^0 \cdot \boldsymbol{C}^2\boldsymbol{D}^0. \tag{11}$$

It is to note that a convenient form of the free-energy function $W(I_i)$ $(i = 1, \ldots, 6)$ is the one where the purely elastic part is split from the remaining part involving coupled electrostatic effects, namely $W(I_i) = W_{elas}(I_1, I_2) + W_{elec}(I_4, I_5, I_6)$.

For the elastic part, two models are the most adopted taking into account the features of soft elastomers: the simple neo-Hookean solid, defined as

$$W_{elas}(I_1) = \frac{\mu}{2}(I_1 - 3), \tag{12}$$

where μ is the shear modulus of the material, and the Gent model, whose strain energy is given by

$$W_{elas}(I_1) = -\frac{\mu J_m}{2}\mathrm{Ln}\left(1 - \frac{I_1 - 3}{J_m}\right). \tag{13}$$

The latter is better suited for rubber-like material at large strains as it incorporates the idea of a limit stretch as the polymer chains extend fully. This limit is defined by the Gent parameter $J_m = I_1^{lim} - 3$ where I_1^{lim} is the first invariant evaluated at the limit stretch. For $J_m \to \infty$, the neo-Hookean model is attained.

When the material behaves as an ideal dielectric, i.e. ε is independent of the strain, a suitable function for W_{elec} is $W_{elec}(I_5)$ in which the coefficient in front of the invariant is a constant, namely $W_{elec} = 1/(2\varepsilon)I_5$. However, sometimes it is necessary to introduce a more involved dependency on the invariants, therefore other expressions of W_{elec} are introduced (see, e.g. Gei et al., 2014) that depends on all the three invariants I_4, I_5, I_6, namely

$$W_{elec} = \frac{1}{2\varepsilon}(\bar{\gamma}_0 I_4 + \bar{\gamma}_1 I_5 + \bar{\gamma}_2 I_6), \tag{14}$$

where $\bar{\gamma}_i$ $(i = 0, 1, 2)$ are dimensionlessconstants such that $\sum_i \bar{\gamma}_i = 1$. As $\bar{\gamma}_0, \bar{\gamma}_2 \to 0$ (and then $\bar{\gamma}_1 \to 1$), the energy function becomes that of an ideal dielectric for which, through Eq. $(9)_2$ and Eqs. (7), the relationship between $\boldsymbol{E}$ and $\boldsymbol{D}$ is

$$\boldsymbol{D} = \varepsilon \boldsymbol{E}. \tag{15}$$

In conclusion of this section it is very important to remark that, as an alternative to the variable $\boldsymbol{D}^0$, electro-elastic constitutive equations can be expressed in terms of electric field $\boldsymbol{E}^0$, so that a free energy $H(\boldsymbol{F}, \boldsymbol{E}^0)$ can be formulated. Taking advantage of the Legendre transform $H(\boldsymbol{F}, \boldsymbol{E}^0) = W(\boldsymbol{F}, \boldsymbol{D}^0) - \boldsymbol{D}^0 \cdot \boldsymbol{E}^0$, thermodynamic arguments provide the following constitutive equations

$$\boldsymbol{S} = \frac{\partial H}{\partial \boldsymbol{F}} - p\boldsymbol{F}^{-T}, \quad \boldsymbol{D}^0 = -\frac{\partial H}{\partial \boldsymbol{E}^0}, \tag{16}$$

that can be adopted in place of (9) (see Dorfmann & Ogden, 2006; Suo et al., 2008 for more details).

A set of electro-elastic invariants similar to (11) may be formulated as a function of the electric field, i.e.

$$\tilde{I}_4 = \boldsymbol{E}^0 \cdot \boldsymbol{E}^0, \quad \tilde{I}_5 = \boldsymbol{E}^0 \cdot \boldsymbol{C}^{-1}\boldsymbol{E}^0, \quad \tilde{I}_6 = \boldsymbol{E}^0 \cdot \boldsymbol{C}^{-2}\boldsymbol{E}^0, \tag{17}$$

so that an electro-elastic neo-Hookean, ideal dielectric follows a free energy

$$H(I_1, \tilde{I}_5) = \frac{\mu}{2}(I_1 - 3) - \frac{\varepsilon}{2}\tilde{I}_5. \tag{18}$$

For the free energy (18) the explicit form of $(16)_1$ turns out to be

$$\boldsymbol{S} + p\,\boldsymbol{F}^{-T} - \mu\,\boldsymbol{F} = \varepsilon\,\boldsymbol{F}^{-T}\,\boldsymbol{E}^0 \otimes \boldsymbol{F}^{-1}\,\boldsymbol{F}^{-T}\,\boldsymbol{E}^0, \tag{19}$$

which corresponds to

$$\boldsymbol{\tau} + p\,\boldsymbol{I} - \mu\,\boldsymbol{F}\,\boldsymbol{F}^T = \varepsilon\,\boldsymbol{E} \otimes \boldsymbol{E}. \tag{20}$$

2.2 Finite Electro-Elastic Actuation of Thin Films

We discuss here, for the same membrane elastomer, three electro-elastic problems, differing from each other by the type of the electric actuation. They are sketched in Fig. 2, that represents the three devices in the deformed configurations. In all cases, the coordinate x_1 is the longitudinal axis of the actuator, x_2 is directed across the thickness and x_3 is the out-of-plane axis. The material is an ideal dielectric and is

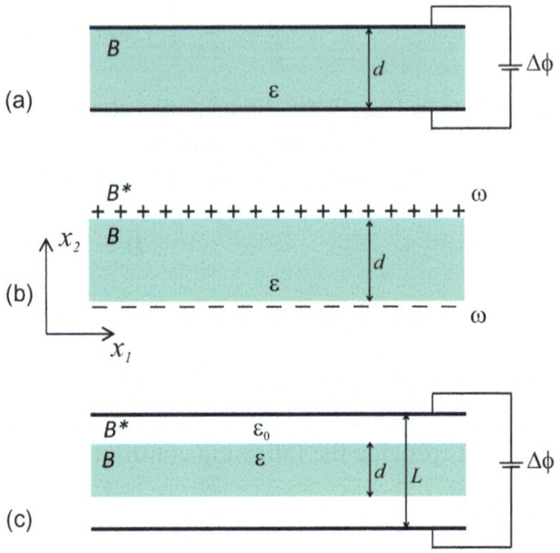

Fig. 2 Sketches of three types of electro-elastic actuation: **a** compliant electrodes; **b** charged-controlled actuation; **c** 'floating' elastomer in vacuum

assumed to be incompressible. The first problem, Fig. 2a, is the classical one where two very compliant electrodes are attached to the two opposite sides of the film (Pelrine et al., 1998). In the second problem, Fig. 2b, charges are laid down directly on the surfaces of the material (sprayed) without the use of electrodes (Keplinger et al., 2010), a method that was first studied experimentally by Röntgen (1880). The third and final problem, Fig. 2c, consists of the elastomer deforming and 'floating' in vacuum between two electrodes held at a fixed distance L (Diaz-Calleja et al., 2009; Liguori & Gei, 2023; Su et al., 2020).

In all cases, the electro-elastic specimen deforms homogeneously in plane-strain from a thickness d_0 to d (fringing effects are neglected), therefore $d = \lambda_2 d_0 = \lambda_1^{-1} d_0$. The electric field is 'transverse' ($\boldsymbol{E} = [0, E_2, 0]$). A relationship between the potential difference $\Delta\phi$ across the device and E_2 can be derived from the fact that in piecewise problems, like those addressed here, the latter is the change in voltage over a given distance. The finite electro-elastic actuation laws of the three problems are summarised below for neo-Hookean elastic, ideal dielectrics.

2.2.1 Compliant Electrodes

The electric field outside the elastomer vanishes and the potential difference across the electrodes is simply

$$\Delta\phi = E_2 d, \tag{21}$$

or, in terms of d_0 and knowing that $D_2 = \varepsilon E_2$ (Eq. 15),

$$\Delta\phi = \frac{d_0 D_2}{\varepsilon \lambda_1}. \tag{22}$$

The neo-Hookean free-energy function (12) is used in Eqs. (9) and (6) to obtain the total stress $\boldsymbol{\tau}$. The boundary conditions are specialised for the case where no electric field is present on the two outer sides of the film: $\tau_{22} = 0$ is used to solve the pressure term p, and $\tau_{11} = 0$ is employed to determine the variable D_2. We can hence obtain the relation for $\Delta\phi$ in terms of λ_1, i.e.

$$\Delta\phi = \frac{d_0\sqrt{\mu\varepsilon^{-1}(\lambda_1^4 - 1)}}{\lambda_1^2}. \tag{23}$$

2.2.2 Charge-Controlled Actuation

When actuation is controlled by the amount of surface charge, whose nominal density is denoted by $\pm\omega_0$, the current longitudinal stretch can be calculated by noting that the jump in the electric displacement, Eq. $(4)_1$, specialises to $D_2 = \omega$. Therefore, for the same free energy, it turns out that

$$\omega = \sqrt{\mu\varepsilon(\lambda_1^2 - \lambda_1^{-2})}. \tag{24}$$

It is worth pointing out that due to the connection between electric field and electric displacement field, the relationship between $\Delta\phi$ and λ_1 coincides with that in Eq. (23).

2.2.3 'Floating' Elastomer in Vacuum

In the third problem, the electric potential jump across the fixed electrodes is provided by

$$\Delta\phi = E_2^*(L - d) + E_2 d, \tag{25}$$

where E_2^* is the only non-vanishing component of the electric field in the vacuum. As the interface between vacuum and elastomer is free from surface charges, from Eq. (4) we know that $\boldsymbol{D}$ is continuous across this interface and hence $D_2^* = D_2$, revealing that the electric displacement field is constant in the space between electrodes. This leads to the equality $\varepsilon_0 E_2^* = \varepsilon E_2$, which helps to achieve an expression for $\Delta\phi$ in terms of E_2 as follows,

$$\Delta\phi = \frac{E_2}{\varepsilon_0}((L - d)\varepsilon + d\varepsilon_0) = \frac{E_2}{\varepsilon_0}((L - d_0\lambda_1^{-1})\varepsilon + d_0\lambda_1^{-1}\varepsilon_0). \tag{26}$$

Equivalently, using Eq. (15), Eq. $(26)_2$ can be expressed in terms of electric displacement D_2, i.e.

$$\Delta\phi = \frac{D_2}{\varepsilon_0\varepsilon}\left((L - d_0\lambda_1^{-1})\varepsilon + d_0\lambda_1^{-1}\varepsilon_0\right). \tag{27}$$

The boundary conditions are specialised for the case where vacuum surrounds the specimen and the Maxwell stress is present $\tau_{22} = T_{22}^*$ determines the pressure term p and $\tau_{11} = T_{11}^*$ yields the electric displacement field D_2. We can hence obtain the relation for $\Delta\phi$ in terms of λ_1

$$\Delta\phi = \frac{(d_0(\varepsilon_0 - \varepsilon) + L\varepsilon\lambda_1)\sqrt{\mu(\lambda_1^4 - 1)}}{\lambda_1^2\sqrt{\varepsilon\varepsilon_0^2 - \varepsilon^2\varepsilon_0}}. \tag{28}$$

3 Linearized Incremental Deformations

This section summarizes the linear incremental deformation theory that allows for the study of the onset of diffuse and localised bifurcation modes (Bertoldi & Gei, 2011).

The way the method works is by superimposing incremental deformations upon a given configuration. Perturbations of nominal surface charges and tractions applied on the boundary of B^0, respectively $\dot{\omega}^0$ and $\dot{\boldsymbol{t}}^0$ are considered such that a new equilibrium configuration is reached where Eqs. (4) and (8) are satisfied. The incremental displacement and deformation gradient are defined similarly to the finite counterpart as $\dot{\boldsymbol{x}} = \dot{\chi}(\boldsymbol{x}^0)$ and $\dot{\boldsymbol{F}} = \text{Grad}\dot{\chi}$, respectively, where a superposed dot $(\dot{\ })$ represents the increment in the quantity concerned. In the current framework, the governing equations (8) turn into the following

$$\text{Div}\dot{\boldsymbol{S}} = \boldsymbol{0}, \quad \text{Div}\dot{\boldsymbol{D}}^0 = 0, \quad \text{Curl}\dot{\boldsymbol{E}}^0 = \boldsymbol{0}. \tag{29}$$

The constitutive equations (9) can be linearised provided that all incremental quantities are small

$$\begin{aligned} \dot{S}_{iJ} &= -\dot{p}F_{Ji}^{-1} + pF_{Li}^{-1}\dot{F}_{kL}F_{Jk}^{-1} + C_{iJkL}^0\dot{F}_{kL} + B_{iJM}^0\dot{D}_M^0, \\ \dot{E}_M^0 &= B_{iJM}^0\dot{F}_{iJ} + A_{MN}^0\dot{D}_N^0, \end{aligned} \tag{30}$$

where $\dot{p}$ is an incremental hydrostatic pressure. The components of the relevant tensors appearing in Eqs. (30) are highlighted to facilitate the understanding between the different quantities. The tensors of electro-elastic moduli $\mathbb{C}^0$, $\mathbb{B}^0$, $\mathbb{A}^0$ can be expressed in terms of the free energy W as

$$C_{iJkL}^0 = \frac{\partial^2 W}{\partial F_{iJ}\partial F_{kL}}, \quad B_{iJM}^0 = \frac{\partial^2 W}{\partial F_{iJ}\partial D_M^0}, \quad A_{MN}^0 = \frac{\partial^2 W}{\partial D_M^0\partial D_N^0}, \tag{31}$$

which lead to the following symmetries

$$C_{iJkL}^0 = C_{kLiJ}^0, \quad A_{MN}^0 = A_{NM}^0.$$

The Lagrangian formulation implied by Eqs. (29) can be turned into an updated Lagrangian one by using push-forward operations based on the new quantities

$$\boldsymbol{\Sigma} = \dot{\boldsymbol{S}}\boldsymbol{F}^T, \quad \hat{\boldsymbol{D}} = \boldsymbol{F}\dot{\boldsymbol{D}}^0, \quad \hat{\boldsymbol{E}} = \boldsymbol{F}^{-T}\dot{\boldsymbol{E}}^0. \tag{32}$$

As a consequence, the updated governing equations take the form

$$\mathrm{div}\boldsymbol{\Sigma} = \boldsymbol{0}, \quad \mathrm{div}\hat{\boldsymbol{D}} = 0, \quad \mathrm{curl}\hat{\boldsymbol{E}} = \boldsymbol{0}. \tag{33}$$

The corresponding incremental boundary conditions can be derived from the boundary conditions given in Eqs. (4) assuming $\boldsymbol{u}(\boldsymbol{x}) = \dot{\boldsymbol{x}}$. In the updated Lagrangian formulation, they are

$$[[\boldsymbol{u}^0]] = \boldsymbol{0}, \quad [[\boldsymbol{\Sigma}]]\boldsymbol{n}dA = \dot{\boldsymbol{t}}^0 dA^0, \quad [[\hat{\boldsymbol{D}}]]\cdot\boldsymbol{n}dA = -\dot{\omega}^0 dA^0, \quad \boldsymbol{n}\times[[\hat{\boldsymbol{E}}]] = \boldsymbol{0}. \tag{34}$$

The boundary conditions can be specialised, likewise Eqs. (5), when the discontinuity is adjacent to vacuum as follows:

$$\boldsymbol{\Sigma}\boldsymbol{n}dA = \dot{\boldsymbol{t}}^0 dA^0 + \boldsymbol{\Sigma}^*\boldsymbol{n}\,dA, \quad \hat{\boldsymbol{D}}\cdot\boldsymbol{n}\,dA = -\dot{\omega}^0 dA^0 + \hat{\boldsymbol{D}}^*\cdot\boldsymbol{n}\,dA, \tag{35}$$

where

$$\boldsymbol{\Sigma}^* = \dot{\boldsymbol{T}}^* + \boldsymbol{T}^*((\mathrm{tr}\boldsymbol{L})\boldsymbol{I} - \boldsymbol{L}^T), \quad \hat{\boldsymbol{D}}^* = \varepsilon_0(\dot{\boldsymbol{E}}^* + ((\mathrm{tr}\boldsymbol{L})\boldsymbol{I} - \boldsymbol{L})\boldsymbol{E}^*). \tag{36}$$

The constitutive equations can be updated through $\boldsymbol{L} = \mathrm{grad}\boldsymbol{u}$, to yield

$$\begin{aligned} \Sigma_{ir} &= C_{irks}L_{ks} + pL_{ri} - \dot{p}\delta_{ir} + B_{irk}\hat{D}_k, \\ \hat{E}_i &= B_{kri}L_{kr} + A_{ik}\hat{D}_k, \end{aligned} \tag{37}$$

where δ_{ij} is the Kronecker delta and the updated electro-elastic moduli become

$$C_{irks} = C^0_{iJkL}F_{rJ}F_{sL}, \quad B_{irk} = B^0_{iJM}F_{rJ}F^{-1}_{Mk}, \quad A_{ik} = A^0_{JM}F^{-1}_{Ji}F^{-1}_{Mk}, \tag{38}$$

with symmetries that apply as before.

In the case the domain outside the solid is vacuum ($B^{\mathrm{sur}} \equiv B^*$), incremental boundary conditions can be stated as shown by Bertoldi and Gei (2011) and Dorfmann and Ogden (2010). Here we consider the case, relevant for practical applications, where both surface tractions $\boldsymbol{t}^0$ and surface charges ω^0 are independent of the deformation (dead loading), thus $\dot{\boldsymbol{t}}^0 = \boldsymbol{0}$ and $\dot{\omega}^0 = 0$, while the electric field in vacuum vanish, as in the space outside a parallel-plate capacitor (by neglecting the edge effects). The consequence is that both the Maxwell stress $\boldsymbol{\tau}^*$ and its increment $\dot{\boldsymbol{\tau}}^*$, the latter being

$$\dot{\boldsymbol{\tau}}^* = \varepsilon_0 \Big[\dot{\boldsymbol{E}}^* \otimes \boldsymbol{E}^* + \boldsymbol{E}^* \otimes \dot{\boldsymbol{E}}^* - (\boldsymbol{E}^* \cdot \dot{\boldsymbol{E}}^*)\, \boldsymbol{I} \Big],$$

vanish, while the increments of $\boldsymbol{D}^*$ and $\boldsymbol{E}^*$ (required in order to satisfy the incremental boundary conditions) are simply related as $\dot{\boldsymbol{D}}^* = \varepsilon_0 \dot{\boldsymbol{E}}^*$. Therefore, also including the incompressibility of the dielectric, the boundary conditions for the Lagrangian formulation of the incremental problem specialize as follows

$$\dot{\boldsymbol{S}}\, \boldsymbol{n}^0 = \boldsymbol{0}, \qquad \dot{\boldsymbol{D}}^0 \cdot \boldsymbol{n}^0 = \varepsilon_0 (\boldsymbol{F}_{\mathrm{b}}^{-1} \dot{\boldsymbol{E}}^*) \cdot \boldsymbol{n}^0, \qquad \boldsymbol{n}^0 \times \dot{\boldsymbol{E}}^0 = \boldsymbol{n}^0 \times \boldsymbol{F}_{\mathrm{b}}^T \dot{\boldsymbol{E}}^*, \tag{39}$$

whereas, with reference to updated Lagrangian variables, they read:

$$\boldsymbol{\Sigma}\, \boldsymbol{n} = \boldsymbol{0}, \qquad \hat{\boldsymbol{D}} \cdot \boldsymbol{n} = \varepsilon_0 \dot{\boldsymbol{E}}^* \cdot \boldsymbol{n}, \qquad \boldsymbol{n} \times \hat{\boldsymbol{E}} = \boldsymbol{n} \times \dot{\boldsymbol{E}}^*. \tag{40}$$

Note that, owing to Eq. $(33)_3$, the incremental electric variable $\dot{\boldsymbol{E}}^*$ in B^* is profitably defined making use of the incremental electrostatic potential in vacuum $\dot{\phi}^*(x_1, x_2)$ as $\dot{E}_i^* = -\dot{\phi}^*_{,i}$; the fulfilment of condition $(33)_2$ in B^* furthermore requires that the potential function is harmonic, i.e.

$$\dot{\phi}^*_{,ii} = 0. \tag{41}$$

4 Global Bifurcations of Soft Dielectric Elastomers

Global bifurcations of the equilibrium paths for a generic electro-elastic system consisting of two media, respectively occupying domains B^0 and $B^{0\mathrm{sur}}$ with reference to a Lagrangian description, can be addressed referring to the general theory introduced by Bertoldi and Gei (2011). Among this class of instabilities, we aim to investigate both electro-mechanical (pull-in) and diffuse-mode bifurcations for an electro-elastic layer, the latter involving the relevant cases of buckling-like and surface-like instabilities.

Along an electro-mechanical loading path, the existence of two distinct solutions of the incremental problem is admitted and the fields generated by their difference, here denoted by symbol Δ (e.g. $\Delta\dot{\chi} = \dot{\chi}^{(1)} - \dot{\chi}^{(2)}$), are taken into account. The difference fields can be regarded as the solution to a homogeneous incremental boundary-value problem (with no associated incremental body forces, tractions, volume free charges, surface charges), thus an extended version of the principle of the virtual work in the Langrangian description requires that

$$\int_{B^0 \cup B^{0\mathrm{sur}}} \left[\Delta\dot{\boldsymbol{S}} \cdot \Delta\dot{\boldsymbol{F}} + \Delta\dot{\boldsymbol{E}}^0 \cdot \Delta\dot{\boldsymbol{D}}^0 \right] dV^0 = 0 \tag{42}$$

for every set of admissible Lagrangian fields $\{\Delta\dot{\chi}, \Delta\dot{\boldsymbol{D}}^0, \Delta\dot{\boldsymbol{S}}, \Delta\dot{\boldsymbol{E}}^0\}$. Note that the existence of the integrals on B^{0sur} requires the decay at infinity of the quantities involved.

Being both $\dot{\boldsymbol{t}}^0$ and $\dot{\omega}^0$ null, the trivial pair $\{\dot{\chi}^{(2)}, \dot{\boldsymbol{D}}^{0(2)}\} = \boldsymbol{0}$ represents a possible solution associated with the incremental boundary-value problem, consequently the difference fields reduce to the solution identified by superscript $^{(1)}$ and Eq. (42) may take the following equivalent form:

$$\int_{B^0 \cup B^{0sur}} \left[\dot{\boldsymbol{S}}^{(1)} \cdot \dot{\boldsymbol{F}}^{(1)} + \dot{\boldsymbol{E}}^{0(1)} \cdot \dot{\boldsymbol{D}}^{0(1)} \right] dV^0 = 0. \tag{43}$$

Therefore, denoting by t ($t \geq 0$) the scalar loading parameter relevant to the principal equilibrium path, a sufficient condition preventing the dielectric layer from the occurrence of a bifurcation is

$$\int_{B^0 \cup B^{0sur}} [\dot{\boldsymbol{S}}(t) \cdot \dot{\boldsymbol{F}} + \dot{\boldsymbol{E}}^0(t) \cdot \dot{\boldsymbol{D}}^0] \, dV^0 > 0, \tag{44}$$

while a bifurcation takes place at $t = t_{cr}$ as soon as, for an admissible critical pair $\{\dot{\chi}_{cr}, \dot{\boldsymbol{D}}^0_{cr}\}$ –that is the *primary eigenmode*–, the functional becomes positive semi-definite. This means that Eq. (43) becomes true, namely

$$\int_{B^0 \cup B^{0sur}} [\dot{\boldsymbol{S}}_{cr}(t_{cr}) \cdot \dot{\boldsymbol{F}}_{cr} + \dot{\boldsymbol{E}}^0_{cr}(t_{cr}) \cdot \dot{\boldsymbol{D}}^0_{cr}] \, dV^0 = 0, \tag{45}$$

with $\dot{\boldsymbol{S}}_{cr}(t_{cr})$ and $\dot{\boldsymbol{E}}^0_{cr}(t_{cr})$ given by the incremental constitutive equations.

The instability criterion defined in Eq. (45) according to the Lagrangian description can be easily given an updated Lagrangian expression through a formal push-forward operation, namely

$$\int_{B \cup B^{sur}} [\boldsymbol{\Sigma}_{cr}(t_{cr}) \cdot \boldsymbol{L}_{cr} + \hat{\boldsymbol{E}}_{cr}(t_{cr}) \cdot \hat{\boldsymbol{D}}_{cr}] \, dV = 0, \tag{46}$$

for an admissible critical pair $\{\boldsymbol{u}_{cr}, \hat{\boldsymbol{D}}_{cr}\}$. Admissibility of $\boldsymbol{u}_{cr}$ and $\hat{\boldsymbol{D}}_{cr}$ requires the fulfilment of field and boundary conditions, namely the incompressibility constraint, div $\boldsymbol{u}_{cr} = 0$, as well as Eqs. $(34)_{1,3}$ and $(33)_2$. Therefore, if $\boldsymbol{\Sigma}_{cr}$ and $\hat{\boldsymbol{E}}_{cr}$ represent the equilibrated incremental total stress and the curl-free incremental electric field (respectively satisfying field equations $(33)_1$ and $(33)_3$) subordinated by $\boldsymbol{u}_{cr}$ and $\hat{\boldsymbol{D}}_{cr}$, the critical condition (46) corresponds to the enforcement of boundary conditions $(34)_{2,4}$ in weak form, as can be easily shown making use of divergence and Stokes' theorems.

When the surrounding medium is vacuum, condition (46) takes the form

$$\int_B [\boldsymbol{\Sigma}_{cr} \cdot \boldsymbol{L}_{cr} + \hat{\boldsymbol{E}}_{cr} \cdot \hat{\boldsymbol{D}}_{cr}] \, dV + \int_{B^*} \boldsymbol{E}^*_{cr} \cdot \boldsymbol{D}^*_{cr} \, dV = 0, \tag{47}$$

which can be simplified as

$$\int_B [\boldsymbol{\Sigma}_{cr} \cdot \boldsymbol{L}_{cr} + \hat{\boldsymbol{E}}_{cr} \cdot \hat{\boldsymbol{D}}_{cr}] \, dV + \varepsilon_0 \int_{\partial B^*} \dot{\phi}^*_{cr} \, \text{grad} \, \dot{\phi}^*_{cr} \cdot \boldsymbol{n} \, dA = 0, \tag{48}$$

through Eq. (41), entailing $\dot{\boldsymbol{E}}^* \cdot \dot{\boldsymbol{D}}^* = \varepsilon_0 \, \text{div}(\dot{\phi}^* \text{grad} \dot{\phi}^*)$, and subsequent application of the divergence theorem to the integral on B^* in (47).

4.1 Electro-Mechanical Instability

This bifurcation may arise when the body is deformed homogeneously as an effect of dead-load tractions/charges applied to its boundary (see the boundary-value problems described in Sect. 2.2), therefore homogeneous perturbation fields $\boldsymbol{L}$, $\hat{\boldsymbol{D}}$ and $\dot{\phi}^*$ are considered. Note that in this case the surface integral in (48) vanishes, being $\text{grad} \, \dot{\phi}^* = \boldsymbol{0}$, therefore, as a result of homogeneity, the instability criterion requires that the argument of the volume integral in (48) vanishes, namely, for an incompressible material,

$$\mathbb{C}(t) \, \boldsymbol{L} \cdot \boldsymbol{L} + p(t) \, \text{tr} \, \boldsymbol{L}^2 + 2\mathsf{B}(t) \, \hat{\boldsymbol{D}} \cdot \boldsymbol{L} + A(t) \, \hat{\boldsymbol{D}} \cdot \hat{\boldsymbol{D}} = 0 \tag{49}$$

for at least a pair $\{\boldsymbol{L}, \hat{\boldsymbol{D}}\} = \{\boldsymbol{L}_{cr}, \hat{\boldsymbol{D}}_{cr}\} \neq \boldsymbol{0}$, with $\text{tr} \, \boldsymbol{L} = 0$.

In conclusion, bifurcation is predicted in correspondence to the loss of positive definiteness of the quadratic form in Eq. (49) (see Gei et al., 2012, for the application of this criterion to a homogeneous actuators and the comparison with the method based on the Hessian of the total energy introduced in Zhao and Suo 2007).

4.2 Diffuse-Mode Instabilities

For an electro-elastic layer, diffuse modes correspond to a plane-strain inhomogeneous response of the specimen with wavelength equal to $2\pi/k_1$ (being k_1 the wave-number of the perturbation). The extreme cases of *long-wavelength, buckling* ($k_1 \to 0$) and *surface instability* ($k_1 \to +\infty$), where the critical modes are strongly localized in the vicinity of the surface, are included. Making reference to Fig. 3, diffuse bifurcation modes can be studied representing the set of admissible incremental fields in condition (48) as sinusoidal functions.

Considering the updated Lagrangian formulation, the incremental boundary-value problem can be written in scalar notation in the form:

$$\Sigma_{11,1} + \Sigma_{12,2} = 0, \quad \Sigma_{21,1} + \Sigma_{22,2} = 0,$$

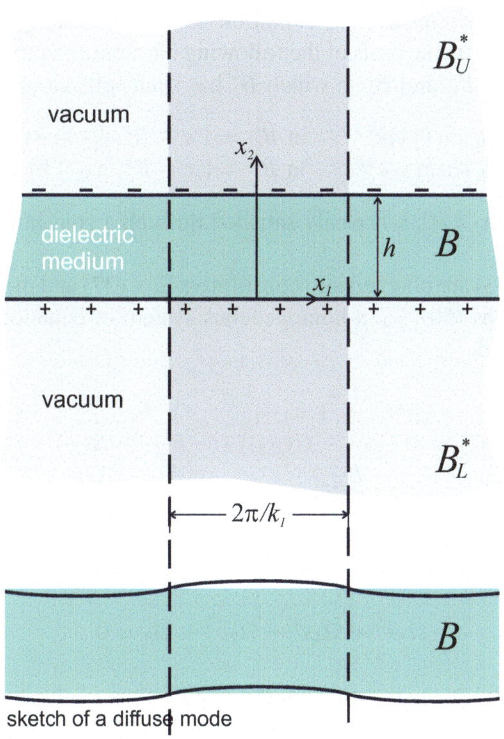

Fig. 3 Sketch of the problem of seeking diffuse-mode bifurcations for a soft electro-elastic layers subjected to a set of charges on its boundaries; in the bottom part, an antisymmetric mode whose wavenumber corresponds to k_1 is depicted

$$\hat{D}_{1,1} + \hat{D}_{2,2} = 0, \quad \hat{E}_{1,2} - \hat{E}_{2,1} = 0 \quad (\text{in } B), \tag{50}$$

$$\dot{E}_i^* = -\dot{\phi}^*_{,i}, \quad \dot{\phi}^*_{,ii} = 0 \qquad (i = 1, 2, \text{ in } B^*), \tag{51}$$

$$\Sigma_{12} = 0, \quad \Sigma_{22} = 0, \quad \hat{D}_2 = \varepsilon_0 \dot{E}_2^*, \quad \hat{E}_1 = \dot{E}_1^* \quad (\text{along } \partial B). \tag{52}$$

The periodic solution adopted inside layer B, namely

$$\begin{aligned} &u_1(x_1, x_2) = V s\, e^{sk_1x_2} \cos k_1x_1, \qquad u_2(x_1, x_2) = V\, e^{sk_1x_2} \sin k_1x_1, \\ &\hat{D}_1(x_1, x_2) = \delta s\, e^{sk_1x_2} \cos k_1x_1, \qquad \hat{D}_2(x_1, x_2) = \delta\, e^{sk_1x_2} \sin k_1x_1, \\ &\dot{p}(x_1, x_2) = Q\, e^{sk_1x_2} \sin k_1x_1, \end{aligned} \tag{53}$$

guarantees that both fields $\boldsymbol{u}$ and $\hat{\boldsymbol{D}}$ are divergence-free, as required by incompressibility and Eq. (50)$_3$ (note that in the case of a compressible dielectric, condition div $\boldsymbol{u} = 0$ should not be enforced, but, simultaneously, variable $\dot{p}$ would disappear).

In order to fulfil the remote decay conditions in the surrounding space, the relevant solution is expressed on the basis of the following harmonic electric potentials inside each of the portions B_U^* and B_L^* in which B^* has been split according to Fig. 3:

- $\dot{\phi}^*(x_1, x_2) = F_U \sin k_1 x_1 e^{-k_1 x_2}$ in $B_U^* = \{\boldsymbol{x} \in B^*,\ x_2 \geq h\}$,
- $\dot{\phi}^*(x_1, x_2) = F_L \sin k_1 x_1 e^{+k_1 x_2}$ in $B_L^* = \{\boldsymbol{x} \in B^*,\ x_2 \leq 0\}$.

The jump $(52)_3$ at $x_2 = 0, h$ is easily satisfied through a convenient choice of constants F_U and F_L.

When modes (53) are plugged into constitutive Eqs. (37) and the resulting expressions into conditions $(50)_{1,2,4}$, a homogeneous system of equations for amplitudes V, δ, Q is generated:

$$\begin{bmatrix} k_1 s(-C_{1111} + C_{1122} + s^2 C_{1212} + C_{1221}) & s^2 B_{121} & -1 \\ -k_1(C_{2121} + s^2(C_{2112} + C_{2211} - C_{2222})) & s(-B_{211} + B_{222}) & -s \\ k_1 s(s^2 B_{121} + B_{211} - B_{222}) & s^2 A_{11} - A_{22} & 0 \end{bmatrix} \begin{bmatrix} V \\ \delta \\ Q \end{bmatrix} = \begin{bmatrix} 0 \\ 0 \\ 0 \end{bmatrix}.$$

Thus a non trivial solution is only admissible when the associated matrix of coefficients is singular, i.e. when the following cubic equation in s^2 is satisfied:

$$\Omega_6 s^6 + \Omega_4 s^4 + \Omega_2 s^2 + \Omega_0 = 0. \tag{54}$$

For the fundamental path investigated, the expressions of the constitutive tensors A_{ik}, B_{irk} and C_{irks} can be calculated from Eqs. (38) and the coefficients of (54) take the form:

$$\begin{aligned} \Omega_6 &= - B_{121}^2 + A_{11} C_{1212}, \\ \Omega_4 &= - 2B_{121}(B_{121} - B_{222}) - A_{22} C_{1212} \\ &\quad - A_{11}(C_{1111} - 2C_{1122} - 2C_{1221} + C_{2222}), \\ \Omega_2 &= - (B_{121} - B_{222})^2 + A_{11} C_{2121} \\ &\quad + A_{22}(C_{1111} - 2C_{1122} - 2C_{1221} + C_{2222}), \\ \Omega_0 &= - A_{22} C_{2121}. \end{aligned} \tag{55}$$

According to the nature of the six solutions s_i, different regimes can be identified and the general solution inside B is built by superposition:

$$
\begin{aligned}
u_1(x_1, x_2) &= \sum_{i=1}^{6} V_i s_i \, e^{s_i k_1 x_2} \cos k_1 x_1, \\
u_2(x_1, x_2) &= \sum_{i=1}^{6} V_i \, e^{s_i k_1 x_2} \sin k_1 x_1, \\
\hat{D}_1(x_1, x_2) &= \sum_{i=1}^{6} \delta_i s_i \, e^{s_i k_1 x_2} \cos k_1 x_1, \\
\hat{D}_2(x_1, x_2) &= \sum_{i=1}^{6} \delta_i \, e^{s_i k_1 x_2} \sin k_1 x_1, \\
\dot{p}(x_1, x_2) &= \sum_{i=1}^{6} Q_i \, e^{s_i k_1 x_2} \sin k_1 x_1.
\end{aligned}
\tag{56}
$$

The critical conditions are now determined by introducing the latter expressions into the stability criterion, Eq. (48), which can be further simplified by taking into account the bounded modular domain highlighted in Fig. 3 as

$$
\int_{-\frac{\pi}{k_1}}^{\frac{\pi}{k_1}} \int_{0}^{h} \left[\boldsymbol{\Sigma} \cdot \boldsymbol{L} + \hat{\boldsymbol{E}} \cdot \hat{\boldsymbol{D}} \right] dx_2 dx_1 +
$$

$$
-\varepsilon_0 \int_{-\frac{\pi}{k_1}}^{\frac{\pi}{k_1}} \left[\dot{\phi}^* \operatorname{grad} \dot{\phi}^* \cdot \boldsymbol{n} \big|_{x_2=h} + \dot{\phi}^* \operatorname{grad} \dot{\phi}^* \cdot \boldsymbol{n} \big|_{x_2=0} \right] dx_1 = 0; \tag{57}
$$

here $\boldsymbol{n}$ denotes the outward normal unit vector relevant to the specific portion of boundary of B (as in Sect. 2).

Equation (57) stems from the remote decay conditions of the electric fields inside vacuum and the periodic nature of the perturbation, allowing the integrals on ∂B^* to vanish along the vertical surfaces bounding the integration domain (corresponding to the dashed lines in Fig. 3). This procedure has been applied to the problem under study and by Gei et al. (2014) it has been verified that the primary eigenmodes so obtained coincide with those evaluated on the basis of the procedure illustrated by Bertoldi and Gei (2011) where all the boundary conditions (34) are enforced in strong form. Details on results when this procedure is applied can be found in Gei et al. (2014).

5 Electro-Mechanics of Laminated Composites Under Plane-Strain Conditions

A method to improve the actuation performance is to create of electro-elastic materials a composite material assembled by embedding a high-dielectric reinforcement particles in a much softer matrix (Huang et al., 2004; Ponte Castañeda & Siboni, 2012; Lopez-Pamies, 2014; Lefevre & Lopez-Pamies, 2017). In a series of papers, it has been shown that a very effective microstructure to achieve a specific performance at equal volume fraction of the component phases is that of hierarchical laminates (deBotton et al., 2007; Tian et al., 2012; Gei et al., 2013; Rudykh et al., 2013; Spinelli & Lopez-Pamies, 2015; Gei & Mutasa, 2018; Marin et al., 2021; Bardella et al., 2022).

We analyse here a layered actuator obtained by repeating a unit cell consisting of two incompressible dielectric phases, a soft matrix with low dielectric constant and a stiff and high-permittivity phase, respectively denoted by 'b' and 'a', with a generic lamination angle θ. Assuming that the length scale of the microstructure is very small compared to the thickness of the actuator, the macroscopic behaviour of the system can be determined on the basis of the homogenization theory (Hill, 1972; Ogden et al., 1974; deBotton, 2005). If h^{0a} and h^{0b} represent the relevant phase thicknesses in the reference, stress-free configuration B^0 (see Fig. 4a), the volume fractions are given by $c^a = h^{0a} / \left(h^{0a} + h^{0b}\right)$ and $c^b = 1 - c^a$, respectively.

The body is deformed by an electro-mechanical loading reaching the current configuration B. In a Lagrangian setting, Eqs. (8) must be satisfied in each phase.

Continuity at all the internal interfaces between phases 'a' and 'b' is enforced by imposing, similarly to (4)

$$[[\boldsymbol{F}]]\, \boldsymbol{m}^0 = \boldsymbol{0}, \quad [[\boldsymbol{S}]]\, \boldsymbol{n}^0 = \boldsymbol{0}, \quad [[\boldsymbol{D}^0]] \cdot \boldsymbol{n}^0 = 0, \quad \boldsymbol{n}^0 \times [[\boldsymbol{E}^0]] = \boldsymbol{0}, \tag{58}$$

where, now, $[[f]] = f^a - f^b$ and $\boldsymbol{n}^0$ is the normal to the interface pointing toward 'b'. In plane strain, continuity of the tangential component of the electric field can be also written as $[[\, \boldsymbol{E}^0]] \cdot \boldsymbol{m}^0 = 0$, where $\boldsymbol{m}^0$ is aligned with the interface and such that $\boldsymbol{n}^0 \cdot \boldsymbol{m}^0 = 0$.

Here the heterogeneous actuator is conceived as a homogenized continuum, therefore its behaviour is described on the basis of the macroscopic, 'average', quantities $\boldsymbol{S}^{\mathrm{av}}$, $\boldsymbol{F}^{\mathrm{av}}$, $\boldsymbol{D}^{0\mathrm{av}}$ and $\boldsymbol{E}^{0\mathrm{av}}$, for which the governing equations (8) are still valid.

For the geometry concerned (Fig. 4a), the electric field outside the actuator vanishes and consequently, denoting by $\tilde{\boldsymbol{n}}^0$ the outward unit normal to the boundaries, the boundary conditions take the expressions

$$\boldsymbol{S}^{\mathrm{av}}\, \tilde{\boldsymbol{n}}^0 = \boldsymbol{0}, \quad \boldsymbol{D}^{0\mathrm{av}} \cdot \tilde{\boldsymbol{n}}^0 = -\omega^0, \quad \tilde{\boldsymbol{n}}^0 \times \boldsymbol{E}^{0\mathrm{av}} = \boldsymbol{0}. \tag{59}$$

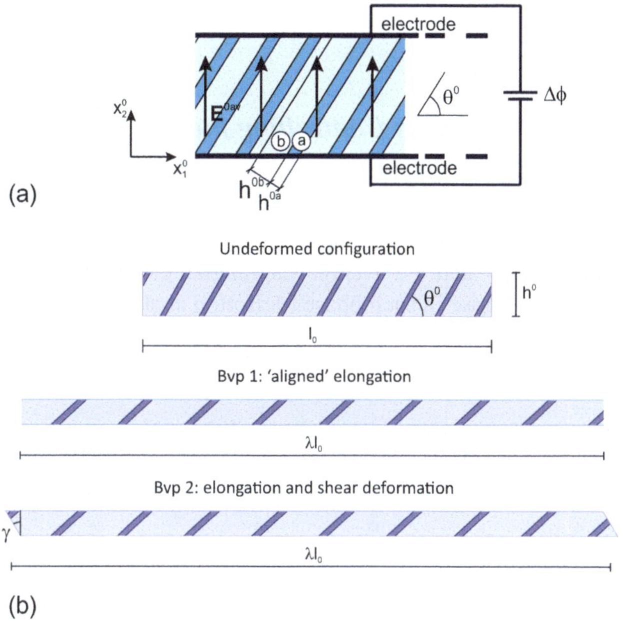

Fig. 4 **a** Sketch of an electro-elastic composite laminate actuator; illustrated by Bertoldi and Gei graphical representations of the boundary-value problems addressed in the section

5.1 Constitutive Assumptions

As already pointed out, the constitutive equations provide the key relationships between the external electric input and the electro-mechanical response of the material. It is assumed that *each phase* of the composite (*i*) is neo-Hookean hyperelastic and (*ii*) behaves as an ideal dielectric, i.e. $\boldsymbol{D} = \varepsilon\, \boldsymbol{E}$. As voltage actuation is investigated, a free energy satisfying the above conditions is that introduced in (18) for which the constitutive equations are (19).

The macroscopic energy of the composite is obtained as the sum of the weighed energies for each phase, namely $H^{\text{av}} = c^a H^a + c^b H^b$, and the macroscopic total stress and electric field can be determined from H^{av} via the homogenized constitutive equations

$$\boldsymbol{S}^{\text{av}} = \frac{\partial H^{\text{av}}}{\partial \boldsymbol{F}^{\text{av}}} - p^{\text{av}}(\boldsymbol{F}^{\text{av}})^{-T}, \qquad \boldsymbol{D}^{0\text{av}} = -\frac{\partial H^{\text{av}}}{\partial \boldsymbol{E}^{0\text{av}}}. \tag{60}$$

We note that, on the basis of the homogenization procedure detailed below, an anisotropic macroscopic behaviour of the laminate is expected, both mechanically and electrically; the latter is evident in the Eulerian constitutive relationship

$\boldsymbol{D}^{\mathrm{av}} = \boldsymbol{\epsilon}^{\mathrm{av}}\, \boldsymbol{E}^{\mathrm{av}}$, where matrix $\boldsymbol{\epsilon}^{\mathrm{av}}$ depends on the properties of both phases, irrespective of the driving actuation.

5.2 Homogenized Solution Controlling the Voltage and Boundary Conditions

Under the assumption of a homogeneous response in each phase, the macroscopic deformation gradient $\boldsymbol{F}^{\mathrm{av}}$ and the average Lagrangian electric displacement $\boldsymbol{E}^{0\mathrm{av}}$ are the weighed sum of those in each phase (deBotton et al., 2007), namely

$$\boldsymbol{F}^{\mathrm{av}} = c^a\, \boldsymbol{F}^a + c^b\, \boldsymbol{F}^b, \qquad \boldsymbol{E}^{0\mathrm{av}} = c^a\, \boldsymbol{E}^{0a} + c^b\, \boldsymbol{E}^{0b}. \tag{61}$$

On the one hand, interface compatibility $(58)_1$ and Eq. $(61)_1$ provide

$$\boldsymbol{F}^a = \boldsymbol{F}^{\mathrm{av}} \left(\boldsymbol{I} + \alpha\, c^b\, \boldsymbol{m}^0 \otimes \boldsymbol{n}^0 \right), \quad \boldsymbol{F}^b = \boldsymbol{F}^{\mathrm{av}} \left(\boldsymbol{I} - \alpha\, c^a\, \boldsymbol{m}^0 \otimes \boldsymbol{n}^0 \right), \tag{62}$$

where α is a real parameter. On the other hand, $(58)_4$ requires that the jump in $\boldsymbol{E}^0$ be along the direction normal to the layers, namely

$$\boldsymbol{E}^{0a} - \boldsymbol{E}^{0b} = \tilde{\beta}\, \boldsymbol{n}^0, \tag{63}$$

where $\tilde{\beta}$ is another real parameter. It follows from $(61)_2$ and (63) that

$$\boldsymbol{E}^{0a} = \boldsymbol{E}^{0\mathrm{av}} + c^b \tilde{\beta}\, \boldsymbol{n}^0, \quad \boldsymbol{E}^{0b} = \boldsymbol{E}^{0\mathrm{av}} - c^a \tilde{\beta}\, \boldsymbol{n}^0.$$

Quantities α and $\tilde{\beta}$ are obtained enforcing $(58)_{2,3}$ after the substitution of the constitutive laws (16). As both phases are described by the free energy (18), their expressions read:

$$\alpha = \frac{\mu^b - \mu^a}{c^a \mu^b + c^b \mu^a} \frac{\boldsymbol{F}^{\mathrm{av}}\, \boldsymbol{n}^0 \cdot \boldsymbol{F}^{\mathrm{av}}\, \boldsymbol{m}^0}{\boldsymbol{F}^{\mathrm{av}}\, \boldsymbol{m}^0 \cdot \boldsymbol{F}^{\mathrm{av}}\, \boldsymbol{m}^0}, \tag{64}$$

$$\tilde{\beta} = \frac{\varepsilon^b - \varepsilon^a}{c^b \varepsilon^a + c^a \varepsilon^b} \frac{(\boldsymbol{F}^{\mathrm{av}})^{-T}\, \boldsymbol{E}^{0\mathrm{av}} \cdot (\boldsymbol{F}^{\mathrm{av}})^{-T}\, \boldsymbol{n}^0}{(\boldsymbol{F}^{\mathrm{av}})^{-T}\, \boldsymbol{n}^0 \cdot (\boldsymbol{F}^{\mathrm{av}})^{-T}\, \boldsymbol{n}^0} + \alpha\, \boldsymbol{E}^{0\mathrm{av}} \cdot \boldsymbol{m}^0. \tag{65}$$

Condition $(58)_2$ can be also used to evaluate the jump of the phase pressures p^a and p^b, yielding

$$p^b - p^a = \left\{ \left[(\boldsymbol{F}^{\mathrm{av}})^{-T}\, \boldsymbol{E}^{0\mathrm{av}} \cdot (\boldsymbol{F}^{\mathrm{av}})^{-T}\, \boldsymbol{n}^0 \right]^2 \frac{\varepsilon^a \varepsilon^b (\varepsilon^a - \varepsilon^b)}{(c^b \varepsilon^a + c^a \varepsilon^b)^2} + \mu^b - \mu^a \right\} \frac{1}{(\boldsymbol{F}^{\mathrm{av}})^{-T}\, \boldsymbol{n}^0 \cdot (\boldsymbol{F}^{\mathrm{av}})^{-T}\, \boldsymbol{n}^0}. \tag{66}$$

The behaviour of the heterogeneous actuator sketched in Fig. 4a is investigated for two different plane-strain boundary-value problems. As typical actuators mainly consist in very thin specimens, in all of them we disregard the edge effects concentrated along the boundary of the electrodes or related to a lack of homogeneity of the surface charge distribution. This means that we assume that the Lagrangian electric field $\boldsymbol{E}^{0\mathrm{av}}$ is directed along x_2^0. Moreover, in all the following problems, we fix the rigid-body motion of the finite deformation such that the two straight boundaries remain aligned with x_1^0 and the electric fields $\boldsymbol{E}^{0\mathrm{av}}$ and $\boldsymbol{E}^{\mathrm{av}} = (\boldsymbol{F}^{\mathrm{av}})^{-T}\,\boldsymbol{E}^{0\mathrm{av}}$ both act along the orthogonal direction, namely

$$\boldsymbol{E}^{0\mathrm{av}} = E^{0\mathrm{av}}\,\boldsymbol{e}_2, \qquad \boldsymbol{E}^{\mathrm{av}} = E^{\mathrm{av}}\,\boldsymbol{e}_2, \tag{67}$$

where $\boldsymbol{e}_2$ is the unit vector directed along x_2^0. The scalar relationship between $E^{0\mathrm{av}}$ and E^{av} depends on the macroscopic deformation gradient $\boldsymbol{F}^{\mathrm{av}}$. In view of the dielectric anisotropy, the electric displacements, $\boldsymbol{D}^{0\mathrm{av}}$ and $\boldsymbol{D}^{\mathrm{av}}$, are not in general aligned with $\boldsymbol{e}_2$. Under voltage-controlled actuation, $E^{0\mathrm{av}} = \Delta\phi/h^0$.

For all the problems addressed in this section the transverse direction remains traction free, namely

$$S_{22}^{\mathrm{av}} = 0, \tag{68}$$

a condition that allows the determination of the unknown pressure p^{av}.

In the first boundary-value problem ('bvp 1', Fig. 4b), we assume that the actuation deform the specimen macroscopically such that the principal directions of strain correspond to x_1^0 and x_2^0 or, alternatively, that $\boldsymbol{F}^{\mathrm{av}}$ admits the diagonal representation $\boldsymbol{F}^{\mathrm{av}} = \mathrm{diag}[\lambda, 1/\lambda, 1]$, where λ is the longitudinal stretch. The electro-mechanical response (i.e., the laws λ–$\Delta\phi$) can be determined imposing null average longitudinal traction, $S_{11}^{\mathrm{av}} = 0$. We note that the average shear stresses are not vanishing in this problem, nevertheless this deformation path is important for two reasons: (*i*) taking into account the slenderness of real devices, it can be concluded that shear deformability does not play a major role in the estimation of the longitudinal stretch at moderate electrical excitation (as it will be clarified in the description of 'bvp 2'); (*ii*) the deformation adopted here is usually employed to investigate micro and macroscopic instabilities of laminates at finite strain (Bertoldi & Gei, 2011; Rudykh et al., 2011; Siboni & Ponte Castañeda, 2014).

In the second problem ('bvp 2'), the device deforms macroscopically at vanishing tractions, namely we impose, in addition to Eq. (68), $S_{11}^{\mathrm{av}} = 0$ and $S_{12}^{\mathrm{av}} = 0$. Due to the intrinsic macroscopic mechanical anisotropy, the admissible deformation gradient is now

$$\boldsymbol{F}^{\mathrm{av}} = \begin{bmatrix} \lambda & \xi/\lambda & 0 \\ 0 & 1/\lambda & 0 \\ 0 & 0 & 1 \end{bmatrix}, \tag{69}$$

where ξ is the amount of shear related to the shear angle γ in the current configuration ($\tan\gamma = \xi$). It is expected that at low voltage/charge the stretch λ will be very close to that computed in 'bvp 1', while diverging at higher electrical excitations.

5.3 Macroscopic Performance

The actuation behaviour of a device subjected to an increasing voltage $\Delta\phi$ is now investigated. The properties of the composite are set by the volume fraction of the soft matrix c^b, the angle of grade of phases in the reference configuration θ^0 and the ratios $m = \mu^a/\mu^b$ and $r = \varepsilon^a/\varepsilon^b$. Results are provided in dimensionless form, so that the scale of voltage is given in terms of the quantity $\overline{\Delta\phi} = (\Delta\phi/h^0)\sqrt{\varepsilon^b/\mu^b}$.

Figure 5 refers to 'bvp 1', where the principal deformations are aligned with axes x_1^0 and x_2^0 (Fig. 4b). The plots show that the performance enhancement of the heterogeneous actuator compared with the homogeneous case (where the device is exclusively composed of the soft polymer 'b') could be remarkable especially at low stretches (i.e. the composite can reach the same stretch λ at lower voltage). The upper limit of the stretch range where this occurs mainly depends on the initial angle θ^0, that is usually chosen in the range [60°, 90°] in order to exploit, along the deformation, the benefit ensuing from the finite rotation of the stiff phase. When the current grade reaches $\theta \approx 45°$ (simple geometrical considerations show that $\tan\theta = \tan\theta^0/\lambda^2$), a strong stiffening in the actuation response takes place, as it becomes more and more difficult to longitudinally stretch the stiff phase beyond this limit. In the figure, θ^0 assumes the values of $\theta^0 = 65°$ and 75°. Even for the most favourable layer grades, when ratios m and r coincide, the performance improvement of the composite is very poor, while it is definitely remarkable when r is one order of magnitude higher than m.

In Fig. 6, the deformation arising in 'bvp 2' is reported. On the one hand, for the longitudinal stretch the same considerations illustrated above for 'bvp 1' still hold true, with an even worse performance of the composite in the case where m and r are coincident. On the other hand, this increased longitudinal stiffness of the heterogeneous material is associated with the occurrence of shear deformations, which evolve very remarkably with the voltage as shown in Fig. 6b: beyond a sort of threshold in voltage, which depends on the ratios m and r of the two phases, the shear angle exhibits large increments for small voltage ranges, growing almost unboundedly. It also appears that the lower the disparity of the two phases involved,

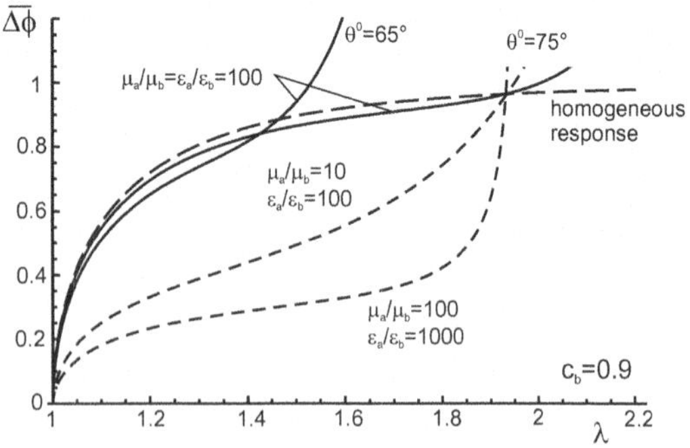

Fig. 5 Response of the laminate for 'bvp 1' ($c_b = 0.9$)

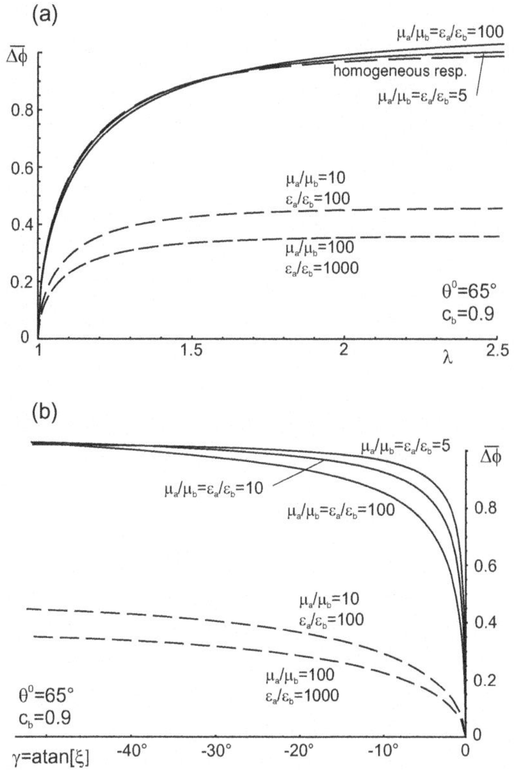

Fig. 6 Response of the laminate for 'bvp 2' ($c_b = 0.9,\ \theta^0 = 65°$) in terms of **a** stretch and **b** shear angle γ

the stiffer is the composite with respect to shearing, tending to the limit case of the homogeneous material, for which $\gamma = 0$, whatever voltage is applied.

In order to conceive actuators based on shear, it is also interesting to investigate the dependence of the attainable shear angles onto the fiber grade, as pictured in Fig. 7. It appears that the behaviour of the composite is strongly influenced by its geometry and, also in this context, angles $\theta^0 \simeq 60°, 70°$ demonstrate to be the most advantageous.

In this section, we have focused on simple laminates, however more efficient performances can be achieved by hierarchical laminates in which the microstructure is constructed by nesting laminates at different order. For more details the reader is referred to, e.g. Lopez-Pamies (2014), Tian et al. (2012), Rudykh et al. (2013), Gei and Mutasa (2018).

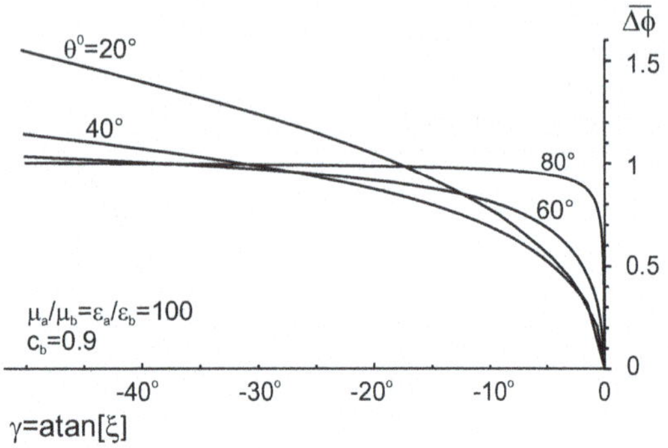

Fig. 7 Response of the laminate for 'bvp 2' ($c_b = 0.9$): shear angle γ as a function of the grade θ^0 of the laminae

6 Introduction to Mechanical-to-Electrical Energy Conversion

A dielectric elastomer generator (DEG) is a highly deformable parallel-plate capacitor made of a soft DE film coated with two compliant electrodes on its opposite faces (Mckay et al., 2011; Kornbluh et al., 2011). The capacitance depends on the deformation undergone by the film (through both the faces area and the thickness), hence it changes during a load-and-release cycle resulting from the interaction of the device with its environment (Fig. 8). This variability can be exploited to extract electric energy by initially stretching, then charging the capacitor and, subsequently, releasing the stretch and collecting the charge at a higher electric potential.

The focus is here on dielectric elastomer generators subjected to a four-stroke electro-mechanical cycle, in which an external oscillating force powers the stretch and contraction cycle. It is assumed that the device deforms under a plane-strain condition that simulates the effect of transverse constraint due to a supporting frame.

By taking into account the properties of the elastomer and the operating conditions dictated by the external environment, with the aid of a constrained optimization algorithm, our goal is to identify those cycles that produce the maximum energy.

This, in turn, reveals the relative role of the various failure mechanisms and provide guidelines for choosing suitable elastic dielectrics for the DEG. Time-dependent

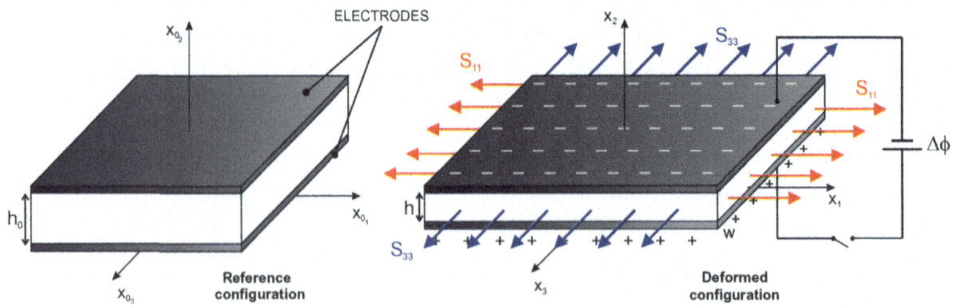

Fig. 8 Sketch of a dielectric elastomer generator in its reference and deformed configurations

effects such as viscosity and loading frequencies are neglected and the results are determined assuming a conservative behaviour of the elastic dielectric.

The generator may undergo different failure mechanisms which must be avoided to ensure proper functioning and long service life of the device. In the assumed plane-strain conditions, the possible limits on the DEG performance are set by the electric breakdown, a state of loss of tension leading to buckling instability, and mechanical failure represented by the maximal threshold of the longitudinal stretch. These failure limits identify a contour of an admissible operational domain that can be depicted in the stress-stretch and the electric potential-charge planes. In both representations the area enclosed within this contour is equal to a theoretical upper bound on the maximal energy that may be harvested from the DEG. The shape of the contour leads us in this work to distinguish between two types of optimal cycles depending on whether or not the electric breakdown limit is attained during the cycle.

6.1 Model of a DE Generator

We consider an ideal dielectric film which is homogeneous, isotropic, hyperelastic and incompressible. The film is stretched from the reference to the deformed configuration by a combination of (*i*) a mechanical force per unit undeformed length s along direction 1 induced by the environment which is the primary source for the energy invested into the system and (*ii*) an electric field generated by an electric potential $\Delta\phi$ across the two stretchable electrodes applied on the opposite surfaces of the film at $x_2 = 0$ and h.

Neglecting fringing effects, the electro-mechanical deformation undergone by the film is homogeneous and can be represented by the deformation gradient $\boldsymbol{F} = \text{diag}[\lambda, \lambda^{-1}, 1]$, where λ is the principal stretch ratio along x_1. Outside the capacitor the electric fields vanish, and the uniform electric field induced by the applied electric potential inside the film is $\boldsymbol{E} = [0, E, 0]^T$, with $E = \Delta\phi/h$. In view of the homogeneous fields developing in the film, the applied force s can be easily related to the total stress $\boldsymbol{S}$, which is divergence free when body forces are null. Thus, along the prescribed loading path $\boldsymbol{S} = \text{diag}[S_{11}, S_{22}, S_{33}]$, where $S_{11} = s/h_0$, $S_{22} = 0$ and S_{33} is the reaction to the kinematic plane-strain constraint. The energy conjugate to the electric field $\boldsymbol{E}$ is the electric displacement field $\boldsymbol{D}$, which is divergence free in the absence of free charges in the material. Within the context of this review it is advantageous to represent the electric fields in terms of their lagrangian counterparts $\boldsymbol{E}^0 = \boldsymbol{F}^T \boldsymbol{E} = [0, E_0, 0]^T$, with $E_0 = E/\lambda$, and $\boldsymbol{D}^0$.

The electro-elastic material is assumed to be governed by a free-energy function $H(\lambda_1, \lambda_2, \lambda_3, E_0)$, where λ_1, λ_2 and λ_3 are the principal stretch ratios such that $\lambda_1\lambda_2\lambda_3 = 1$. The constitutive equations are analogous of (16) and adapted to the principal directions of stretch, thus they read

$$S_{ii} = \frac{\partial H}{\partial \lambda_i} - p\frac{1}{\lambda_i}, \qquad D_0 = -\frac{\partial H}{\partial E_0}, \tag{70}$$

In particular, the form (18) of $H(\lambda_1, \lambda_2, \lambda_3, E_0)$ is adopted that can be written in terms of principal stretches as

$$H = \frac{\mu}{2}(\lambda_1^2 + \lambda_2^2 + \lambda_3^2 - 3) - \frac{\varepsilon}{2}\left(\frac{E_0}{\lambda_2}\right)^2. \tag{71}$$

In the sequel we find advantageous to rephrase the equations in terms of the dimensionless variables

$$\bar{S}_{11} = \frac{S_{11}}{\mu}, \quad \bar{S}_{33} = \frac{S_{33}}{\mu}, \quad \bar{\phi} = \frac{\Delta\phi}{h_0}\sqrt{\frac{\varepsilon}{\mu}}, \quad \bar{\omega}_0 = \frac{\omega_0}{\sqrt{\varepsilon\mu}}. \tag{72}$$

Next, with the aid of Eq. (71) the components of the applied stress can be related to λ and $\Delta\phi$. Accordingly, during the harvesting cycle the relations between the applied stress, the applied electric potential, the resulting stretch ratio and the charge accumulated on the electrodes are:

$$\bar{S}_{11} = \lambda - \frac{1}{\lambda^3} - \bar{\phi}^2\lambda, \quad \bar{S}_{33} = 1 - \frac{1}{\lambda^2} - \bar{\phi}^2\lambda^2, \quad \bar{\phi} = \frac{\bar{\omega}_0}{\lambda^2}. \tag{73}$$

6.2 *The Load-Driven Harvesting Cycle and the Area of Admissible Configurations*

A few possible harvesting strategies, which are primarily distinguished according to the control parameters chosen for the four-stroke cycles, are discussed in Lallart et al. (2012). Here we focus on the cycle illustrated in Fig. 9: along two of the strokes the longitudinal stress S_{11} due to the external load is fixed, whereas along the other two strokes the device is electrically isolated and the charge ω_0 in the electrodes is held fixed. The order of these strokes and their role in the harvesting cycle are:

- mechanical loading A–B: the work produced by the external oscillating force during its rise from minimal to maximal values is stored in the elastically stretching generator. In terms of the dimensionless variables, $\bar{S}_{11}$ increases from its minimal value at A to its maximal value at B, while the charge on the electrodes is fixed ($\bar{\omega}_0^A = \bar{\omega}_0^B$). Due to the stretching of the film the capacitance of the DEG increases and the electric potential drops;
- electrical charging B–C: the electrodes are charged by means of an external service battery such that the electric potential between them is $\Delta\bar{\phi} = \bar{\phi}_C - \bar{\phi}_B > 0$. Along this step the stress is held constant ($\bar{S}_{11}^B = \bar{S}_{11}^C$). Thanks to the attraction between the two charged electrodes the film further shrinks in the x_2-direction and elongates in the x_1-direction. When this stroke terminates the film attains the largest stretch ratio during the cycle (λ_C);
- mechanical unloading C–D: during the decline of the external force from its maximal to minimal values the film shrinks while the charge in the isolated electrodes

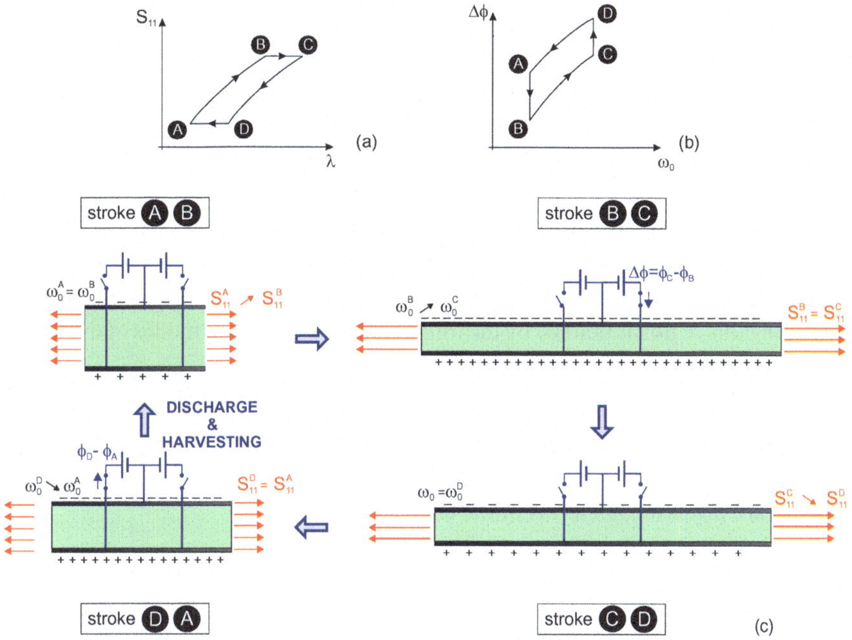

Fig. 9 **a** The harvesting cycle plotted on the mechanical plane; **b** the same on the electrical plane; **c** characterization of the four strokes with a service battery at the right and a storage battery at the left

is fixed ($\bar{\omega}_0^C = \bar{\omega}_0^D$). The thickening of the shrinking film results in an increase of the electric potential to $\bar{\phi}_D$, which is the largest value of $\bar{\phi}$ along the cycle;

- electrical energy-harvesting D–A: the charge deposited in stroke B–C is recollected at a higher electric potential with an appropriate electrical circuit. This energy-harvesting stroke is executed with a fixed load ($\bar{S}_{11}^D = \bar{S}_{11}^A$), nonetheless, due to the decrease in the electric potential between the electrodes and hence the associate decrease in the attracting force between them, the film further shrinks.

The cycle can be presented on the planes illustrated in Fig. 9a and b, with the former corresponding to the mechanical S_{11}–λ plane and the latter to the electrical $\Delta\phi$–ω_0 plane. In passing, we note that in practice strokes B–C and D–A are substantially shorter than the mechanical loading and unloading strokes. Thus, the applied external force should be considered as a continuously oscillating force, so that, when it attains its maximal and minimal values, appropriate electrical circuits are temporarily connected to the electrodes.

The four-stroke harvesting cycle described above is characterized by the four equalities presented in Fig. 9c. These induce the following relations among the corresponding eight dimensionless independent variables:

$$\begin{aligned}
\bar{\omega}_0^A = \bar{\omega}_0^B &\Rightarrow \bar{\phi}_A \lambda_A^2 = \bar{\phi}_B \lambda_B^2,\\
\bar{S}_{11}^B = \bar{S}_{11}^C &\Rightarrow \lambda_B - \frac{1}{\lambda_B^3} - \bar{\phi}_B^2 \lambda_B = \lambda_C - \frac{1}{\lambda_C^3} - \bar{\phi}_C^2 \lambda_C,\\
\bar{\omega}_0^C = \bar{\omega}_0^D &\Rightarrow \bar{\phi}_C \lambda_C^2 = \bar{\phi}_D \lambda_D^2,\\
\bar{S}_{11}^D = \bar{S}_{11}^A &\Rightarrow \lambda_D - \frac{1}{\lambda_D^3} - \bar{\phi}_D^2 \lambda_D = \lambda_A - \frac{1}{\lambda_A^3} - \bar{\phi}_A^2 \lambda_A.
\end{aligned} \tag{74}$$

In order to ensure a proper operational condition of the device, all feasible cycles must lie inside the region of admissible states for the generator (Fig. 10). The line enveloping this region is defined by the following possible failure modes of the DEG:

- electric breakdown (EB): this failure occurs when the electric field E reaches the dielectric strength of the material E_{eb}. In dimensionless form the dielectric strength is $\bar{E}_{eb} = E_{eb}\sqrt{\varepsilon/\mu}$. The corresponding portions of the failure envelops surrounding the region of admissible states in the mechanical and the electric planes are prescribed in terms of the curves:

$$\bar{S}_{11} = \lambda - \frac{1}{\lambda^3} - \frac{\bar{E}_{eb}^2}{\lambda}, \quad \bar{\phi} = \frac{\bar{E}_{eb}^2}{\bar{\omega}_0}; \tag{75}$$

- ultimate stretch (λ_U): this failure takes place when the magnitude of the stretch attains a critical value λ_U at which mechanical failure initiates. The curves that correspond to this failure mode in the mechanical and the electric planes are, respectively:

$$\lambda = \lambda_U, \quad \bar{\phi} = \frac{\bar{\omega}_0}{\lambda_U^2}; \tag{76}$$

- loss of tension (S_{33}=0): to avoid failure due to buckling or wrinkling in compression (Gei et al., 2014) it is required that the two in-plane stresses be positive. This failure mode is associated with the geometrical configuration of the device and is related to the small thickness of the film. A comparison between Eqs. $(73)_1$ and $(73)_2$ for

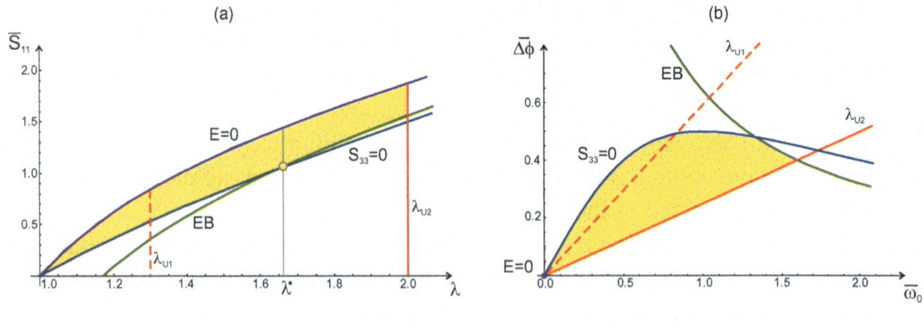

Fig. 10 Regions of admissible states (highlighted in yellow) in the mechanical (**a**) and electrical (**b**) planes

the two stresses reveals that the inequality $S_{33} \geq 0$ is more restrictive than $S_{11} \geq 0$. Therefore, by manipulating these expressions it is found that the portions of the failure envelops corresponding to loss of tension along the x_3-direction in the two pertinent planes are characterized by the curves

$$\bar{S}_{11} = \lambda - \frac{1}{\lambda}, \quad \bar{\phi} = \frac{\bar{\omega}_0}{1 + \bar{\omega}_0^2}. \tag{77}$$

We finally add a fourth formal condition ($E = 0$), which is not related to a failure of the system, requiring that the direction of the electric field is not reversed during the cycle, i.e. $E \geq 0$.

6.3 *Optimization of the Harvesting Cycle*

According to the four-stroke cycle described in Sect. 6.2, the energy-density generated per unit shear modulus is

$$K_g = \frac{\tilde{K}_g}{\mu V_0} = \int_B^C \bar{\phi}\, d\bar{\omega}_0 + \int_D^A \bar{\phi}\, d\bar{\omega}_0. \tag{78}$$

The sum of the two integrals at the right-hand-side of Eq. (78) is equal to the area bounded by the cycle in the electrical plane in its dimensionless form. For later reference we recall that, since K_g is an energy extracted from the system, its sign is negative.

Evaluating the integrals and expressing the constant nominal stresses $\bar{S}_{11}^B$ and $\bar{S}_{11}^A$ in terms of the relevant stretches and of the dimensionless nominal electric potentials via Eq. $(73)_1$, the explicit expression for the dimensionless harvested energy turns to

$$K_g = \frac{1}{2}(\lambda_A - \lambda_D)\left[\lambda_D\left(3\bar{\phi}_D^2 - 1\right) + 2\lambda_A + 3\lambda_D^{-3}\right]$$
$$+ \frac{1}{2}(\lambda_C - \lambda_B)\left[\lambda_B\left(3\bar{\phi}_B^2 - 1\right) + 2\lambda_C + 3\lambda_B^{-3}\right]. \tag{79}$$

Note that the expression for K_g involves only the squares of the variables $\bar{\phi}_B$ and $\bar{\phi}_D$. Therefore, it is convenient to derive $\bar{\phi}_B^2$ and $\bar{\phi}_D^2$ from conditions (74) as functions of the stretches λ_A, λ_B, λ_C and λ_D and substitute them in (79). This leads to an expression for K_g in terms of the four characteristic stretches. Similar developments can be followed for the constraints defining the failure envelope.

In order to determine the optimal cycle through which the maximum energy can be harvested while keeping it within the region of admissible states, we formulate the following constrained optimization problem:

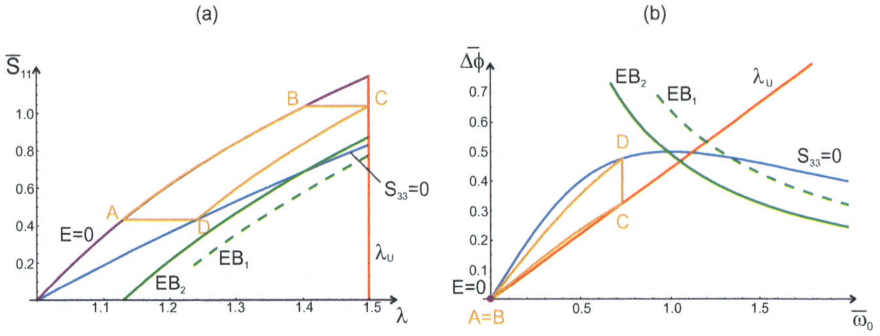

Fig. 11 Examples of an optimal cycle ABCD in the mechanical (**a**) and electrical (**b**) planes

$$\text{find} \min_{\boldsymbol{\Lambda}} K_g[\lambda_A, \lambda_B, \lambda_C, \lambda_D],$$

with $\boldsymbol{\Lambda} = [\lambda_A, \lambda_B, \lambda_C, \lambda_D]^T$ and the minimum is sought since $K_g \leq 0$. The optimization is to be evaluated under the following constraints:

- on the stretches

$$\lambda_C = \lambda_U, \quad 1 \leq \lambda_A \leq \lambda_U, \quad 1 \leq \lambda_B \leq \lambda_U, \quad 1 \leq \lambda_D \leq \lambda_U; \tag{80}$$

  the constraint $(80)_1$ enabled us to substitute λ_U for λ_C reducing the set of optimization variables $\boldsymbol{\Lambda}$ to $\boldsymbol{\Lambda}_R = [\lambda_A, \lambda_B, \lambda_D]^T$;
- on the D state

$$S_{33}^D[\lambda_A, \lambda_B, \lambda_D] \geq 0, \quad \bar{E}_D[\lambda_A, \lambda_B, \lambda_D] \leq \bar{E}_{eb}^2; \tag{81}$$

  these constraints require the compatibility of the D state with the failure modes due to loss of tension and electric breakdown.

An example is reported graphically in Fig. 11. More details on this problem can be found in Springhetti et al. (2014).

7 Conclusions

Over the past twenty years, the theory of electro-elasticity has proven to be a powerful tool to investigate soft dielectric devices and predict their behaviour even though real materials are subjected to –sometimes relevant– dissipative effects. In this chapter, we review the main features of both non-linear and linearized formulations, the latter useful to seek bifurcations.

Equilibrium paths for typical types of actuation of thin dielectric films are first analyzed, then the incremental theory is presented as a prelude for the theoretical framework to determine global bifurcations of soft dielectric elastomers that is applied to study both electro-mechanical instability and diffuse-mode instabilities (which comprise buckling and surface instability).

In the second part, electro-elasticity is applied to obtain homogenized constitutive equations for simple dielectric laminates to be able, on the one hand, to solve some relevant boundary-value problems, on the other, to assess their capability to enhance actuation with respect to simple homogeneous materials. The analysis of the properties of a four-stroke dielectric energy converter, that is subjected to several failure modes, concludes the chapter.

References

Bardella, L., Volpini, V., & Gei, M. (2022). On the effect of the volumetric deformation in soft dielectric composites with high phase contrast. *Journal of Elasticity, 148*, 167–198.

Bertoldi, K., & Gei, M. (2011). Instability in multilayered soft dielectrics. *Journal of the Mechanics and Physics of Solids, 59*, 18–42.

Bortot, E., Denzer, R., Menzel, A., & Gei, M. (2014). Analysis of a viscous soft dielectric elastomer generator operating in an electric circuit. *International Journal of Solids and Structures, 78–79*, 205–215.

Calabrese, L., Berardo, A., De Rossi, D., Gei, M., Pugno, N. M., & Fantoni, G. (2019). A soft robot structure with limbless resonant, stick and slip locomotion. *Smart Materials and Structures, 28*, 104005.

Calabrese, L., Frediani, G., Gei, M., De Rossi, D., & Carpi, F. (2018). Active compression bandage made of dielectric elastomers. *IEEE/ASME Transactions on Mechatronics, 23*, 2328–2337.

Carpi, F. (Ed.). (2016). *Electromechanically active polymers*. Springer.

Carpi, F., Frediani, G., Gerboni, C., Germignani, J., & De Rossi, D. (2014). Enabling variable-stiffness hand rehabilitation orthoses with dielectric elastomer transducers. *Medical Engineering & Physics, 36*, 205–211.

Chen, Y., Zhao, H., Mao, J., Chirarattananon, P., Helbling, E. F., Hyun, N.-S.P., Clarke, D. R., & Wood, R. J. (2019). Controlled flight of a microrobot powered by soft artificial muscles. *Nature, 575*, 324–329.

Conn, A. T., & Rossiter, J. M. (2012). Towards holonomic electro-elastomer actuators with six degrees of freedom. *Smart Materials and Structures, 21*, 035012.

deBotton, G. (2005). Transversely isotropic sequentially laminated composites in finite elasticity. *Journal of the Mechanics and Physics of Solids, 53*, 1334–1361.

deBotton, G., Tevet-Deree, L., & Socolsky, E. A. (2007). Electroactive heterogeneous polymers: Analysis and applications to laminated composites. *Mechanics of Advanced Materials and Structures, 14*, 13–22.

Diaz-Calleja, R., Sanchis, M. J., & Riande, E. (2009). Effect of an electric field on the bifurcation of a biaxially stretched incompressible slab rubber. *European Physical Journal E, 30*, 417–426.

Dorfmann, A., & Ogden, R. W. (2005). Nonlinear electroelasticity. *Acta Mechanica, 174*, 167–183.

Dorfmann, A., & Ogden, R. W. (2006). Nonlinear electroelastic deformations. *Journal of Elasticity, 82*, 99–127.

Dorfmann, A., & Ogden, R. W. (2010). Nonlinear electroelastostatics: Incremental equations and stability. *International Journal of Engineering Science, 48*, 1–14.

Gei, M., Colonnelli, S., & Springhetti, R. (2012). A framework to investigate instabilities of homogeneous and composite dielectric elastomer actuators. *Proceedings of SPIE, 8340*, 834010.

Gei, M., Springhetti, R., & Bortot, E. (2013). Performance of soft dielectric laminated composites. *Smart Materials and Structures, 22*, 104014.

Gei, M., Colonnelli, S., & Springhetti, R. (2014). The role of electrostriction on the stability of dielectric elastomer actuators. *International Journal of Solids and Structures, 51*, 848–860.

Gei, M., & Mutasa, K. C. K. (2018). Optimisation of hierarchical dielectric elastomer laminated composites. *International Journal of Non-Linear Mechanics, 103*, 266–273.

Gupta, U., Qin, L., Wang, Y., Godaba, H., & Zhu, J. (2019). Soft robots based on dielectric elastomer actuators: a review. *Smart Materials and Structures, 28*, 103002.

Hill, R. (1972). On constitutive macro-variables for heterogeneous solids at finte strain. *Proceedings of the Royal Society A, 326*, 131–147.

Huang, C., Zhang, Q. M., deBotton, G., & Bhattacharya, K. (2004). All-organic dielectric-percolative three-component composite materials with high electromechanical response. *Applied Physics Letters, 84*, 4391–4393.

Keplinger, C., Kaltenbrunner, M., Nikita, A., & Bauer, S. (2010). Röntgen's electrode-free elastomer actuators without electromechanical pull-in instability. *PNAS, 107*, 4505–4510.

Kornbluh, R. D., Pelrine, R., Prahlad, H., Wong-Foy, A., McCoy, B., Kim, S., Eckerle, J., & Low, T. (2011). From boots to buoys: promises and challenges of dielectric elastomer energy harvesting. *Proceedings of SPIE, 7976*, 797605.

Lallart, M., Cottinet, P. J., Guyomar, D., & Lebrun, L. (2012). Electrostrictive polymers for mechanical energy harvesting. *Journal of Polymer Science Part B: Polymer Physics, 50*, 523–535.

Lefevre, V., & Lopez-Pamies, O. (2017). Nonlinear electroelastic deformations of dielectric elastomer composites: I-Ideal elastic dielectrics. *Journal of the Mechanics and Physics of Solids, 99*, 409–437.

Liguori, G., & Gei, M. (2023). Surface instabilities of soft dielectric elastomers with implementation of electrode stiffness. *Mathematics & Mechanics of Solids, 28*, 479–500.

Lopez-Pamies, O. (2014). Elastic dielectric composites: Theory and application to particle-filled ideal dielectrics. *Journal of the Mechanics and Physics of Solids, 64*, 61–82.

Lu, T., Ma, C., & Wang, T. (2020). Mechanics of dielectric elastomer structures: A review. *Extreme Mechanics Letters, 38*, 100752.

Lu, T., Shi, Z., Shi, Q., & Wang, T. (2016). Bioinspired bicipital muscle with fiber constrained dielectric elastomer actuator. *Extreme Mechanics Letters, 6*, 75–81.

Marin, F., Martinez-Frutos, J., Ortigosa, R., & Gil, A. J. (2021). A convex multi-variable based computational framework for multilayered electro-active polymers. *Computer Methods in Applied Mechanics and Engineering, 374*, 113567.

McGough, K., Ahmed, S., Frecker, M., & Ounaies, Z. (2014). Finite element analysis and validation of dielectric elastomer actuators used for active origami. *Smart Materials and Structures, 23*, 094002.

McKay, T. G., O'Brien, B. M., Calius, E. P., & Anderson, I. A. (2011). Soft generators using dielectric elastomers. *Applied Physics Letters, 98*, 142903.

McMeeking, R. M., & Landis, C. M. (2005). Electrostatic forces and stored energy for deformable dielectric materials. *Journal of Applied Mechanics, Transactions ASME, 72*, 581–590.

Moretti, G., Rosati Papini, G. P., Daniele, L., Forehand, D., Ingram, D., Vertechy, R., & Fontana, M. (2019). Modelling and testing of a wave energy converter based on dielectric elastomer generators. *Proceedings of the Royal Society A, 475*, 20180566.

Ogden, R. W. (1974). On the overall moduli of non-linear elastic composite materials. *Journal of the Mechanics and Physics of Solids, 22*, 541–553.

Pelrine, R., Kornbluh, R. D., & Joseph, J. (1998). Electrostriction of polymer dielectrics with compliant electrodes as a means of actuation. *Sensors Actuators A, 64*, 77–85.

Pelrine, R., Kornbluh, R. D., Pei, Q., & Joseph, J. (2000). High-speed electrically actuated elastomers with strain greater than 100%. *Science, 287*, 836–839.

Ponte Castañeda, P., & Siboni, M. N. (2012). A finite-strain constitutive theory for electroactive polymer composites via homogenization. *International Journal of Non-Linear Mechanics, 47*, 293–306.
Röntgen, W. C. (1880). Ueber die durch Electricität bewirkten Form-und Volumeänderungen von dielectrischen Körpern. *Annals of Physics, 247*, 771–786.
Rudykh, S., Lewinstein, A., Uner, G., & deBotton, G. (2013). Analysis of microstructural induced enhancement of electromechanical coupling in soft dielectrics. *Applied Physics Letters, 102*, 151905.
Rudykh, S., & deBotton, G. (2011). Stability of anisotropic electroactive polymers with application to layered media. *ZAMP, 62*, 1131–1142.
Siboni, M. N., & Ponte Castañeda, P. (2014). Fiber-constrained, dielectric-elastomer composites: Finite-strain response and stability analysis. *Journal of the Mechanics and Physics of Solids, 68*, 211–238.
Spinelli, A. S., & Lopez-Pamies, O. (2015). Some simple explicit results for the elastic dielectric properties and stability of layered composites. *International Journal of Engineering Science, 88*, 15–28.
Springhetti, R., Bortot, E., deBotton, G., & Gei, M. (2014). Optimal energy-harvesting cycles for load-driven dielectric generators in plane strain. *IMA Journal of Applied Mathematics, 79*, 929–946.
Su, Y., Chen, W., Dorfmann, L., & Destrade, M. (2020). The effect of an exterior electric field on the instability of dielectric plates. *Proceedings of the Royal Society A, 476*, 20200267.
Suo, Z., Zhao, X., & Greene, W. H. (2008). A nonlinear field theory of deformable dielectrics. *Journal of the Mechanics and Physics of Solids, 56*, 467–486.
Tian, L., Tevet-Deree, L., deBotton, G., & Bhattacharya, K. (2012). Dielectric elastomer composites. *Journal of the Mechanics and Physics of Solids, 60*, 181–198.
Wingert, A., Lichter, M. D., & Dubowsky, S. (2006). On the design of large degree-of-freedom digital mechatronic devices based on bistable dielectric elastomer actuators. *IEEE/ASME Transactions on Mechatronics, 11*, 448–456.
Zhao, X., & Suo, Z. (2007). Method to analyze electromechanical stability of dielectric elastomers. *Applied Physics Letters, 91*, 061921.

A Unified Theoretical Modeling Framework for Soft and Hard Magnetorheological Elastomers

Kostas Danas

Abstract These notes put together a number of theoretical and numerical models and results obtained for magnetically soft and hard magnetorheological elastomers, denoted as s-MREs and h-MREs, respectively over the last five years in our group. We present in a unified manner both s- and h-MREs. In particular, we regard MREs, in the general case, as magnetically dissipative nonlinear elastic composite materials comprising a mechanically-soft, non-magnetic elastomeric matrix in which mechanically-rigid, magnetically-dissipative particles are embedded isotropically and randomly. The proposed incremental variational frameworks are general enough to deal with more complex microstructures such as particle-chains or others that do not yet exist in the lab. More importantly, we propose homogenization-guided, analytical, explicit models that are consistent as one moves from the dissipative h-MREs to the purely energetic s-MREs. In parallel, we propose numerical frameworks allowing to simulate a very wide variety of microstructures and boundary value problems in magneto-mechanics.

1 Introduction

By now a large body of research exists in the literature on magneto-active solids. There are mainly two classes of magneto-active solids:

(i) mechanically-stiff magnetic materials made of permanents *hard* magnets (e.g., NdFeB in polycrystal or powder form embedded in a stiff polymer or at very large volume fractions) exhibiting significant magnetic dissipation or magnetically *soft* ferromagnets (e.g., Fe and compounds of it) with negligible magnetic dissipation.

K. Danas (✉)
LMS, C.N.R.S, École Polytechnique, Institut Polytechnique de Paris, 91128 Palaiseau, France
e-mail: konstantinos.danas@polytechnique.edu

K. Danas and O. Lopez-Pamies (eds.), *Electro- and Magneto-Mechanics of Soft Solids*,
CISM International Centre for Mechanical Sciences 610,
https://doi.org/10.1007/978-3-031-48351-6_3

(ii) mechanically-soft polymer-based composite materials (e.g. with shear modulus less than 1 MPa) comprising the former magnetic metals in powder or micron-sized particle form.

The main difference between those two classes of magneto-active materials is the ability of the latter soft polymeric ones to deform substantially under the application of a magnetic field. Usually in the first class of mechanically-stiff magneto-active solids, one may reach *magnetostrictive* strains ranging from a few to several thousands ppm (i.e., in the order of 10^{-6}–10^{-3}). By contrast, use of magneto-active polymers also known as "magnetorheological elastomers" or MREs[1] allows to obtain fairly large effective magnetostrictive strains even as large as 0.4–0.5 (depending on the compliance of the polymer), which is several orders of magnitude larger than those of metals or stiff polymers.

In turn, the first class of magnetic materials exhibits superior magnetic properties (such as initial magnetic permeability or susceptibility, magnetic saturation, magnetic coercivity and so on) as compared to the corresponding soft polymer composites. The reason for this is fairly simple and is related to the actual quantity of particles that can be embedded in the polymer retaining at the same time its soft mechanical properties. In practice, most of mechanically-soft MREs contain volume fractions of magnetizable particles ranging between 0 and 30vol%. Any further addition of rigid particles in the polymer leads rapidly to significant material stiffening and thus to loss of its interesting magnetostrictive capabilities.

The deformation mechanisms in those two classes of materials are entirely different and require in general very different modeling approaches. While one can write down the same general concepts and equations relating magnetics with mechanics, the details required to describe adequately these two classes of materials are extremely different. The metallic materials deform under the application of an externally applied magnetic field due to a complex motion, interaction, nucleation, annihilation and reconfiguration of magnetic walls lying at the nanometer scale. Several nano-, micro- and macroscopic theories have been proposed to describe those deformation mechanisms over the last fifty years and will not be discussed in the present notes. In turn, MREs, which is the main topic of the present note, deform due to rearrangement and (static/dynamic) reconfiguration of the magnetic particles, which carry along with them the surrounding non-magnetic soft polymer matrix. In this sense, the word *magnetostriction* in MREs serves to describe this collective particle rearrangement leading to an effective (or average) deformation of the composite polymer. Evidently, the magnetic particles themselves also deform due to magnetic domain wall motions at the lower nanometer scale. Nonetheless, those strains are orders of magnitude smaller than those occurring at the larger scale of the polymer

[1] In the literature, one may find a number of equivalent names such as MAEs (magneto-active elastomers), MAPs (magneto-active polymers), MREs (magnetorheological elastomers), MSEs (magneto-sensitive elastomers) and many others. The word MREs was the first one, perhaps because these materials were first synthesized by material scientists (Jolly et al., 1996; Ginder et al., 1999) that worked before on magnetorheological fluids, a close cousin to the present mechanically-soft MREs.

composite (several tens of microns all the way up to millimeter scale) and thus are usually neglected to simplify the overall analysis.

While the above discussion focused on the material response, recent works, in the last decade, have used and synthesized MREs in simple or more complex macroscopic shapes, such as beams, films, spheres, cylinders, ellipsoids, trusses or even more fancier shapes mainly with the advent of 3D-printing capabilities. The literature is too vast to include in this introductory section, but interested readers may refer to numerous reviews on the subject in the literature (see for instance Bastola and Hossain (2020) and Lucarini et al. (2022b)). Precisely, this manufacturing capability to obtain complex geometrical, structural shapes made from MREs has led to their rapid development and more recent use in high end biomedical and sensing applications. By contrast, the original metallic magnetic materials are extremely difficult to shape (especially NdFeB or cobalt) since they are very brittle and thus their use is constrained to larger devices or engines (such as electric motors) where complex shaping is not needed at least up to now.

Having made this distinction, it is worth mentioning that in the literature, it is usually common to mix the notions of a material and a structure into the combined word of *meta-material.* Even so, whether those polymer composites are fabricated using magnetically soft particles, denoted henceforth as s-MREs, or magnetically hard particles, denoted as h-MREs, they exhibit extremely interesting coupled magneto-mechanical properties both at the material and structural level and for that reason they are still under extensive investigation.

In these notes, we focus on the modeling of MREs both at the material and the structural level. We present first the general finite strain theory on magnetically dissipative magneto-active solids and focus on the modeling of isotropic (quasi)-incompressible MREs. Subsequently, those material models are assessed by full three-dimensional representative volume element (RVE) finite element (FE) periodic simulations. A special augmented potential energy is developed for the results to be sensible. The developed models are based on homogenization theory and numerical data but also comprise phenomenological parts. This last choice is done in order to obtain simple, robust and above all explicit analytical material models for the MREs. These models may then be implemented in general purpose finite element codes (such as ABAQUS (2017) or FEniCS). Their simplicity allows to model complex two-dimensional and three-dimensional boundary value problems, including "meta-material" (or more correctly meta-structural) response of MRE-based structures, such as homogeneously or inhomogeneously magnetized beams or films.

2 Preliminary Definitions in Magneto-Elasto-Statics

In this section, we lay out some important definitions required for the development of general finite strain models for MREs. We state at this point that the subsequent presentation of the theory of MREs lies naturally at the domain of continuum mechanics and at scales that are that of the particles and larger. With this, we make clear that the goal of the following quantities and equations is to describe the response of MREs at a scale much larger (i.e. several micro-meters and above) than that of the so-called magnetic domain walls, which naturally occur at a scale of a few nano-meters. The reason for these approximations is two-fold. First, the main interest is to provide simple and explicit constitutive laws for composite MRE materials at the macroscale (i.e. millimeter and above) bringing from the lower scale the effect of magnetic particles embedded in a soft polymer matrix. Second, the nucleation and evolution of domain walls usually occurs rather fast and at the scale of the particles, which in turn strongly interact with each other. It is therefore evident that it is extremely difficult (if not practically impossible) to resolve the domain-wall scale together with the particle interaction scale and at the same time recover a simple explicit model that can serve to analyze real boundary value problems (BVPs). For these reasons, we choose to describe the magnetic quantities in the particles in a rather continuum way, i.e., by assuming that pointwise (i.e. at $\mathbf{X}$) at each particle the magnetic quantities are described by continuum magnetic vector variables that follow simple constitutive laws (albeit nonlinear and dissipative ones) without however resorting to the resolution of the domain walls. It is noted that a gradient finite-strain variational phase-field framework following the early works of James and Kinderlehrer (1993) and DeSimone and James (2002) has been proposed recently by Keip and Sridhar (2018). In that work, the authors managed to resolve domain wall evolution at the level of the particles and then numerically resolve a small assembly of them in two-dimensions. They have shown qualitatively and quantitatively similar responses with those obtained by ignoring those domain walls and simply using local constitutive magnetic laws at the particle level. We thus follow the second option in the present note.

2.1 Finite Strain Kinematics

For simplicity, we consider at this stage a deformable, magneto-active solid with volume $\mathcal{V}_0$ ($\mathcal{V}$) in its reference (current) configuration that is embedded in $\mathbb{R}^3$. The domain $\mathbb{R}^3 \backslash \mathcal{V}_0$ is in general non-magnetizable and usually serves to denote the surrounding air (see Fig. 1). The magneto-active solid may not necessarily be a homogeneous body but could comprise several magnetic (e.g., particles) or non-magnetic phases (e.g. polymer matrix). For simplicity, at this introductory stage of the article, the reader could regard the body as homogeneous to avoid confusion for the various definitions.

The boundary of the solid is assumed to be smooth and is designated by $\partial\mathcal{V}_0$ ($\partial\mathcal{V}$), while $\boldsymbol{\mathcal{N}}$ ($\mathbf{n}$) denotes the unit normal on $\partial\mathcal{V}_0$ ($\partial\mathcal{V}$) in the reference (current) configuration (see Fig. 1). The deformation of the solid from the reference to current configuration is defined to be a continuous, twice differentiable (except on the boundary/interfaces), one-to-one mapping $\mathbf{y}(\mathbf{X})$. Thus, the position of any point $\mathbf{X}$ in the reference configuration is given by $\mathbf{x} = \mathbf{y}(\mathbf{X})$ in the current configuration. In

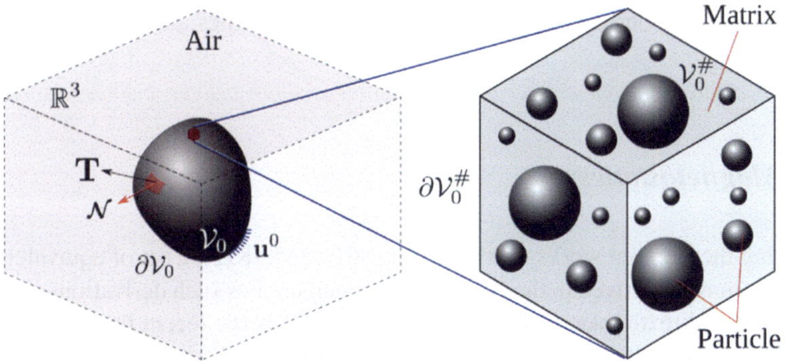

Fig. 1 (Left) Typical sketch of the reference configuration of a magneto-active solid of volume $\mathcal{V}_0$ and boundary $\partial\mathcal{V}_0$ (usually in the order of several mm-cm) embedded in an air domain, occupying a volume $\mathbb{R}^3\backslash\mathcal{V}_0$. (Right) Sketch of a representative material unit-cell comprising usually a non-magnetic polymer matrix phase and mechanically-stiff magnetizable particles with sizes ranging from 2–50 μm

turn, the mechanical displacement field $\mathbf{u}(\mathbf{X})$ relates the current position vector to the reference so that $\mathbf{x} = \mathbf{X} + \mathbf{u}(\mathbf{X})$. The deformation gradient is then defined to be

$$\mathbf{F} = \text{Grad}\, \mathbf{y}(\mathbf{X}) = \mathbf{I} + \text{Grad}\, \mathbf{u}(\mathbf{X}), \qquad J = \det \mathbf{F} > 0, \tag{1}$$

where $\mathbf{I}$ is the second-order identity tensor, whereas $J > 0$ imposes the impenetrability condition, or simply positive volume condition. It is recalled that the Grad $= \partial/\partial\mathbf{X}$ as well as that $\mathbf{F}$ is a compatible two-point tensor, i.e., satisfies Curl$\mathbf{F} = \mathbf{0}$.

After introducing the conjugate quantity to the deformation $\mathbf{F}$, the first Piola-Kirchhoff stress $\mathbf{S}$, conservation and angular momentum lead to the pointwise equilibrium, symmetry and boundary conditions

$$\text{Div}\, \mathbf{S} = \mathbf{0} \quad \text{in} \quad \mathbb{R}^3, \qquad \mathbf{S}\mathbf{F}^T = \mathbf{F}\mathbf{S}^T, \qquad [\![\mathbf{S}]\!] \cdot \mathcal{N} - \mathbf{T} = \mathbf{0} \quad \text{on} \quad \partial\mathcal{V}_0\backslash\partial\mathcal{V}_0^{\mathbf{u}}, \tag{2}$$

where $\partial\mathcal{V}_0^{\mathbf{u}}$ and $\mathbf{T}$ denote the displacement Dirichlet part of the solid boundary and the mechanical traction in the reference configuration. One may write the equivalent form of the equilibrium, symmetry and boundary conditions in the current configuration by introducing the Cauchy stress $\boldsymbol{\sigma}$, such that

$$\text{div}\, \boldsymbol{\sigma} = \mathbf{0} \quad \text{in} \quad \mathbb{R}^3, \qquad \boldsymbol{\sigma} = \boldsymbol{\sigma}^T, \qquad [\![\boldsymbol{\sigma}]\!] \cdot \mathbf{n} - \mathbf{t} = \mathbf{0} \quad \text{on} \quad \partial\mathcal{V}\backslash\partial\mathcal{V}^{\mathbf{u}}. \tag{3}$$

Here, $\partial\mathcal{V}^{\mathbf{u}}$ and $\mathbf{t}$ denote the displacement Dirichlet part of the solid boundary and the mechanical traction in the current configuration, respectively. One may show using a push forward or pull-back transformation that (Ogden, 1997)

$$\boldsymbol{\sigma} = \frac{1}{J}\mathbf{S}\mathbf{F}^T, \qquad \text{or } S = J\boldsymbol{\sigma}\mathbf{F}^{-T}. \tag{4}$$

The above definitions are standard in continuum mechanics and will not be further analyzed.

2.2 *Magnetostatics*

Following the seminal works of Maxwell (1861,1865,1873), a set of equivalent balance laws may be derived in the context of magnetism. For such derivations, one may refer to several textbooks or thesis documents, such as the recent thesis manuscript of Mukherjee (2020). We proceed by briefly recalling those laws and definitions. Traditionally, in the physics community and in the absence of electric currents and charges, one may use three quantities to describe the magnetic state of a solid in the *current* configuration:

- the magnetic field **b**,
- the h-field **h**,
- the magnetization **m** which is naturally zero in non-magnetic domains.

It is important to note, however, that those three quantities are not independent of each other. They are related by the *constitutive* relation

$$\mathbf{b} = \mu_0(\mathbf{h} + \mathbf{m}) \qquad \text{or} \qquad \mathbf{m} = \frac{1}{\mu_0}\mathbf{b} - \mathbf{h} \quad \text{in} \;\; \mathcal{V}, \tag{5}$$

where μ_0 is the magnetic permeability of vaccuum, air or non-magnetic solids.

Remark 2.1 In fact, the second expression in (5) is a *definition* of the magnetization vector in the current volume $\mathcal{V}$, which however is not defined on its boundary $\partial\mathcal{V}$. By definition $\mathbf{m} = \mathbf{0}$ in a non-magnetic body. This implies that as a quantity is insufficient to describe the presence of magnetic lines (in the sense of Maxwell) in the surrounding air or in a non-magnetic solid such as the polymer matrix which is also of interest here. In these two last cases, one is left with the relation $\mathbf{b} = \mu_0\mathbf{h}$, which implies–in the sense of a continuum medium–that **b** is linearly dependent on **h** and *vice versa* via the constitutive parameter μ_0. Henceforth, we will focus on the original Maxwell fields **b** and **h** as independent variables that are related via linear and/or nonlinear constitutive laws, while **m** will serve as a quantity that can be readily computed by Eq. (5), but instead will not be considered as an "independent" variable itself. In fact, in Sect. 7.1 we show by example that **m** may be regarded as an internal state variable in most of the cases.

2.2.1 Ampère's Law

`Assumptions.` The present study pertains to magneto-mechanical problems ignoring all electrical effects. This suggests that, henceforth, we consider that any effects resulting from electric charges, electric fields and varying electric displacements are negligible. Moreover, this implies that any variations of magnetic fields should be sufficiently small such that $\partial \mathbf{b}/\partial t$ is negligible. Thus, we do not deal with high frequency effects in the present study. Moreover for further simplicity, we assume that there exist no surface currents $\mathbf{k}$ throughout this work and there are no relativistic effects in the matter.

`Current configuration.` In the general case of free volume currents, $\mathbf{j}$, running across a surface $\partial\mathcal{V}$ in the current configuration, but in the absence of electric charges and time varying electric displacements (i.e., $\partial \mathbf{d}/\partial t$) and surface currents (known as the Eddy current approximation), Ampère's law takes the following form (dropping for simplicity the explicit dependence on $\mathbf{x}$):

$$\oint_C \mathbf{h} \cdot \mathrm{d}\boldsymbol{\ell} = \int_{\partial\mathcal{V}} \mathbf{j} \cdot \mathbf{n} \, \mathrm{d}\mathcal{S}. \tag{6}$$

In this equation, C is a smooth arbitrary closed curve lying on the smooth closed surface $\partial\mathcal{V}$, $\mathrm{d}\boldsymbol{\ell}$ is an infinitesimal line element with direction tangent to the curve C, and $\mathbf{n}$ is the outward normal to $\partial\mathcal{V}$. By assuming sufficient smoothness of $\mathbf{h}$, use of the standard Stokes theorem leads to the differential form of (6), which reads

$$\int_{\partial\mathcal{V}} \operatorname{curl} \mathbf{h} \cdot \mathbf{n} \, \mathrm{d}\mathcal{S} = \int_{\partial\mathcal{V}} \mathbf{j} \cdot \mathbf{n} \, \mathrm{d}\mathcal{S}, \tag{7}$$

where the operator curl is defined with respect to the deformed position $\mathbf{x}$.

In order to retrieve seamlessly the boundary conditions along a given interface, one may generalize this last result to the case of a surface $\partial\mathcal{V}$ containing a hypothetical *line of discontinuity* (an argument that will be subsequently dropped) γ where the corresponding integrand $\operatorname{curl} \mathbf{h}$ is not defined. In that case, by use of the modified Stoke's theorem (Eringen, 1967; Eringen & Maugin, 1990), the last equation may be rewritten as

$$\int_{\partial\mathcal{V}\setminus\gamma} \operatorname{curl} \mathbf{h} \cdot \mathbf{n} \, \mathrm{d}\mathcal{S} + \oint_\gamma [[\mathbf{h}]] \cdot \mathrm{d}\boldsymbol{\ell} = \int_{\partial\mathcal{V}\setminus\gamma} \mathbf{j} \cdot \mathbf{n} \, \mathrm{d}\mathcal{S}, \tag{8}$$

where continuity of $\mathbf{j}$ is considered along the surface $\partial\mathcal{V}$.

These last integrals together with the arbitrary nature of the surface S and the localization lemma lead readily to the pointwise differential equations and jump conditions (Eringen & Maugin, 1990)

$$\operatorname{curl}\mathbf{h} = \mathbf{j} \quad \text{in} \quad \mathcal{V}, \qquad [\![\mathbf{h}]\!] \times \mathbf{n} = \mathbf{0} \quad \text{on} \quad \partial\mathcal{V}. \tag{9}$$

In the absence of free currents, i.e., $\mathbf{j} = \mathbf{0}$, Ampère's law (9) reads

$$\operatorname{curl}\mathbf{h} = \mathbf{0} \quad \text{in} \quad \mathcal{V}, \qquad [\![\mathbf{h}]\!] \times \mathbf{n} = \mathbf{0} \quad \text{on} \quad \partial\mathcal{V}. \tag{10}$$

The last relation also holds in the air domain which may extend to $\mathbb{R}^3\backslash\mathcal{V}$ or in non-magnetic domains enclosed in $\mathcal{V}$. Moreover, this last form will be used in most of this study, while (9) has been used in the study of Dorn et al. (2021) but is not included in the present notes.

`Reference configuration.` In an exactly similar fashion, one may rewrite the above laws in the reference configuration. This is a necessary operation in the present work, since we deal with finite strains and it is natural to use Lagrangian measures. Thus, in the general case of free volume currents, $\mathbf{J}$, running across a surface $\partial\mathcal{V}_0$ in the reference configuration and in the absence of surface currents $\mathbf{K}$), electric charges and time varying electric displacements (i.e., $\partial\mathbf{D}/\partial t$), Ampère's law takes the following form (using also the standard Stoke's theorem):

$$\oint_{C_0} \mathbf{h}(\mathbf{X}) \cdot \mathbf{F}\mathrm{d}\boldsymbol{\ell}_0 = \int_{\partial\mathcal{V}_0} \mathbf{j} \cdot J\mathbf{F}^{-T}\boldsymbol{\mathcal{N}}\,\mathrm{d}\mathcal{S}_0 \qquad \text{or}$$

$$\int_{\partial\mathcal{V}_0} \operatorname{Curl}\mathbf{H} \cdot \boldsymbol{\mathcal{N}}\,\mathrm{d}\mathcal{S}_0 = \int_{\partial\mathcal{V}_0} \mathbf{J} \cdot \boldsymbol{\mathcal{N}}\,\mathrm{d}\mathcal{S}. \tag{11}$$

Here, C_0 is a smooth arbitrary closed curve lying on the smooth closed surface $\partial\mathcal{V}_0$, $\mathrm{d}\boldsymbol{\ell}_0$ is an infinitesimal line element with direction tangent to the curve C_0, $\boldsymbol{\mathcal{N}}$ is the outward normal to $\partial\mathcal{V}_0$ and the operator Curl is defined with respect to the undeformed position $\mathbf{X}$. Moreover, we have used the Lagrangian definitions

$$\mathbf{H} = \mathbf{F}^T\,\mathbf{h}, \quad \mathbf{J} = J\,\mathbf{F}^{-1}\mathbf{j}, \qquad \text{or} \qquad \mathbf{h} = \mathbf{F}^{-T}\,\mathbf{H}, \quad \mathbf{j} = J^{-1}\,\mathbf{F}\mathbf{J}, \tag{12}$$

as well as the standard line and surface (i.e., Nahnson formula) transformations between the current and reference configuration

$$\mathrm{d}\boldsymbol{\ell} = \mathbf{F}\mathrm{d}\boldsymbol{\ell}_0, \qquad \mathbf{n}\mathrm{d}\mathcal{S} = J\mathbf{F}^{-T}\boldsymbol{\mathcal{N}}\mathrm{d}\mathcal{S}_0. \tag{13}$$

Again considering a hypothetical *line of discontinuity* (an argument that will be subsequently dropped) Γ where the integrand Curl$\mathbf{H}$ may not be defined, the second integral equation in (11) can be written as

$$\int_{\partial\mathcal{V}_0\backslash\Gamma} \operatorname{Curl}\mathbf{H} \cdot \boldsymbol{\mathcal{N}}\,\mathrm{d}\mathcal{S}_0 + \oint_{C\backslash\Gamma} [\![\mathbf{H}]\!] \times \boldsymbol{\mathcal{N}}\,\mathrm{d}\boldsymbol{\ell}_0 = \int_{\partial\mathcal{V}_0} \mathbf{J} \cdot \boldsymbol{\mathcal{N}}\,\mathrm{d}\mathcal{S}. \tag{14}$$

Given the arbitrariness of the surface $\partial\mathcal{V}_0$ and curves C_0 and Γ, the above operations readily lead by use of the localization lemma to the pointwise differential equations

$$\mathrm{Curl}\,\mathbf{H} = \mathbf{J} \quad \text{in} \quad \mathcal{V}_0, \qquad [[\mathbf{H}]] \times \boldsymbol{\mathcal{N}} = \mathbf{0} \quad \text{on} \quad \partial\mathcal{V}_0. \tag{15}$$

In the absence of free currents, i.e., $\mathbf{J} = \mathbf{0}$, Ampère's law (9) reads

$$\mathrm{Curl}\,\mathbf{H} = \mathbf{0} \quad \text{in} \quad \mathcal{V}_0, \qquad [[\mathbf{H}]] \times \boldsymbol{\mathcal{N}} = \mathbf{0} \quad \text{in} \quad \partial\mathcal{V}_0. \tag{16}$$

The last relation also holds in the air domain which may extend to $\mathbb{R}^3 \backslash \mathcal{V}_0$ or in non-magnetic domains enclosed in $\mathcal{V}_0$.

Definition 2.2 (*Scalar potential*) In the absence of free currents, i.e., when $\mathbf{J} = \mathbf{0}$, one may define a scalar potential $\varphi(\mathbf{X})$ such that

$$\mathbf{H}(\mathbf{X}) = -\mathrm{Grad}\varphi(\mathbf{X}) \tag{17}$$

which implies

$$\mathrm{Curl}(\mathrm{Grad}\varphi) = \mathbf{0} \quad \text{in} \quad \mathcal{V}_0 \qquad [[\varphi]] = 0 \quad \text{in} \quad \partial\mathcal{V}_0.$$

This definition allows to satisfy identically relation $(16)_1$, since the curl of a gradient of a scalar field is identically zero. This definition is extremely useful since it allows to satisfy one of the two Maxwell equations (see the second equation in the next section). This condition is similar to the one in mechanics, where the deformation gradient is given in terms of the gradient of a displacement vector field and in that case one satisfies identically the compatibility condition $\mathrm{Curl}\mathbf{F} = \mathbf{0}$.

It is noted here that one could define in a similar fashion a scalar potential in the current configuration. Nonetheless, in the present work, all variables will be defined in the reference configuration and thus such a definition is not of use. Furthermore, in a small strain setting the reference and current configurations coincide and again such a distinction becomes inconsequential.

2.2.2 Absence of Magnetic Monopole—Faraday's Law

`Current configuration.` Under the same assumptions stated in the previous section and under the hypothesis of the absence of magnetic monopole, the normal component of a sufficiently smooth magnetic field $\mathbf{b}(\mathbf{x})$ integrated over a closed smooth surface $\partial\mathcal{V}$ vanishes identically, such that

$$\int_{\partial\mathcal{V}} \mathbf{b} \cdot \mathbf{n}\, \mathrm{d}\mathcal{S} = 0. \tag{18}$$

Use of the modified divergence theorem (Eringen, 1967) considering potential jumps across a discontinuous surface s leads to

$$\int_{\mathcal{V}\backslash s} \text{div}\mathbf{b}\, \partial\mathcal{V} + \int_s [[\mathbf{b}]] \cdot \mathbf{n}\, \text{d}\mathcal{S} = 0. \tag{19}$$

Using the localization lemma and the arbitrariness of the volume $\mathcal{V}$ and the surfaces $\partial\mathcal{V}$ and s, we obtain the general pointwise balance equations and boundary conditions

$$\text{div}\mathbf{b} = 0 \quad \text{in} \quad \mathcal{V}, \qquad [[\mathbf{b}]] \cdot \mathbf{n} = 0 \quad \text{on} \quad \partial\mathcal{V}. \tag{20}$$

Again, this last relation also holds in the air domain which may extend to $\mathbb{R}^3\backslash\mathcal{V}$ or in non-magnetic domains enclosed in $\mathcal{V}$.

`Reference configuration.` Similarly, one may obtain the corresponding balance laws in the reference configuration by using the transformations (13) in (18) and the modified divergence theorem to write

$$\int_{\partial\mathcal{V}_0} \mathbf{b} \cdot J\mathbf{F}^{-T}\mathcal{N}\, \text{d}\mathcal{S}_0 = 0 \quad \Rightarrow \quad \int_{\partial\mathcal{V}_0\backslash\Sigma} \text{Div}\mathbf{B}\, \partial\mathcal{V}_0 + \int_\Sigma [[\mathbf{B}]] \cdot \mathcal{N}\, \text{d}\mathcal{S}_0 = 0, \tag{21}$$

Here, again, $\partial\mathcal{V}_0$ is a closed surface in the undeformed configuration, Σ a discontinuous surface, while the operator Div is defined with respect to the undeformed position $\mathbf{X}$. Moreover, we have used the definitions

$$\mathbf{B} = J\,\mathbf{F}^{-1}\mathbf{b}, \qquad \text{or} \qquad \mathbf{b} = J^{-1}\,\mathbf{F}\mathbf{B}. \tag{22}$$

Using again the fundamental lemma of the variational calculus for the arbitrariness of the $\mathcal{V}_0$ and $\partial\mathcal{V}_0$, we obtain the pointwise balance equations and boundary conditions

$$\text{Div}\mathbf{B} = 0 \quad \text{in} \quad \mathcal{V}_0, \qquad [[\mathbf{B}]] \cdot \mathcal{N} = 0 \quad \text{in} \quad \partial\mathcal{V}_0. \tag{23}$$

Again, this last relation also holds in the air domain which may extend to $\mathbb{R}^3\backslash\mathcal{V}_0$ or in non-magnetic domains enclosed in $\mathcal{V}_0$.

Definition 2.3 (*Vector potential*) One may *always* define a vector potential $\mathbf{A}(\mathbf{X})$ such that

$$\mathbf{B}(\mathbf{X}) = \text{Curl}\mathbf{A}(\mathbf{X}), \tag{24}$$

which implies

$$\text{Div}(\text{Curl}\mathbf{A}) = 0 \quad \text{in} \quad \mathcal{V}_0, \qquad [[\mathbf{A}]] \times \mathcal{N} = 0 \quad \text{in} \quad \partial\mathcal{V}_0.$$

This definition allows to satisfy identically relation $(23)_1$, since the divergence of a curl of a vector field is identically zero. This condition is similar to the one in mechanics, where the stress field may be written in terms of an Airy stress function thus allowing to satisfy identically the equilibrium equations $\mathrm{Div}\mathbf{S} = \mathbf{0}$ (in the absence of body forces and inertial terms).

Again, it is noted here that one could define in a similar fashion a vector potential in the current configuration. Nonetheless, in the present work, all variables will be defined in the reference configuration and thus such a definition is not of use. In turn, in a small strain setting the reference and current configurations coincide and such a distinction becomes inconsequential.

For simplicity, we can summarize the main transformation rules that will be extensively used in the following between Eulerian and Lagrangian quantities as

$$\sigma = \frac{1}{J}\mathbf{S}\mathbf{F}^T, \qquad \mathbf{b} = \frac{1}{J}\mathbf{F}\mathbf{B} \quad \text{and} \quad \mathbf{h} = \mathbf{F}^{-T}\mathbf{H}. \tag{25}$$

3 Thermodynamics and General Variational Formulations

In this section, we present a general incremental thermodynamically consistent variational framework allowing to describe in a general manner dissipative and nonlinear magnetoelastic solids. The presentation that follows is carried out such that stationarity of the variational formulations leads to the previously defined balance laws and boundary conditions. Moreover, one of the pilars of the proposed framework is the Generalized Standard Materials (GSM) approach (Halphen & Nguyen, 1975). Therein, we assume that there exists an energy density W and a dissipation potential D that depend respectively on the chosen independent variables and their rates. The subsequent formulations include therefore the balance laws and Maxwell equations, the boundary conditions and the magneto-mechanical constitutive material laws. This is, in our opinion, the most elegant and concise manner to describe the magneto-mechanical solids of interest here. On the contrary, they are not absolutely necessary in the sense that the set of derived equations are those used in practice to carry out the simulations. Note, however, that writing the problem in a minimum energy principle allows in some cases of interest to obtain rigorous results in the context of instabilities.

Remark 3.1 Henceforth, we propose two equivalent formulations the **F**–**H** and the **F**–**B** one, and as their name states clearly the first considers the magnetic scalar potential and the second the magnetic vector potential as independent magnetic variables, respectively. An alternative formulation may use an additional variable, **m** (Danas et al., 2012; Danas & Triantafyllidis, 2014). Such formulation still is equivalent to the first two (see for example the appendix in Danas (2017), or Bustamante et al. (2008) and more recently Sharma and Saxena (2020)). Nevertheless, **m** is not a sufficiently general variable but instead is defined via the Eq. (5). The main reason for this is that one cannot use only **m** and apply magnetic boundary conditions, since it is not defined at the boundary (or at interfaces) as opposed to **h** or **b** (see discussion in Sect. 3). Even so, in some problems involving magnetic domains and permanently magnetized solids that the state of magnetization is given in terms of a *known* amplitude and a direction locally, it might be advantageous to work directly with $\mathbf{m}$ to obtain analytical and numerical solutions (James & Kinderlehrer, 1993; DeSimone & James, 2002; Keip & Sridhar, 2018). In the present case, however, since we are interested in solving general boundary value problems, one must retain either φ or $\mathbf{A}$ to analyze the magnetic fields in non-magnetic domains (e.g., matrix or air), where the magnetization is null. Thus the addition of $\mathbf{m}$ does not offer any clear advantage, instead it requires additional numerical approximations (such as static condensation in order to deal with jumps across interfaces).

3.1 Scalar Potential-Based F-H Formulation

In this formulation, we consider the mechanical deformation $\mathbf{u}$ and the scalar magnetic potential φ to be the independent primary variables along with a set of internal variables $\boldsymbol{\xi}$.[2] For this to be valid, we need to assume also that we have no free currents, i.e., $\mathbf{J} = \mathbf{0}$. The case of free currents can only be considered in the $\mathbf{F}$–$\mathbf{B}$ setting presented in the following section.

Following the statements in Sect. 2.2, one may define the rate of the potential energy stored in the system, shown in Fig. 1, as

$$\dot{\mathscr{P}}^{\mathrm{H}} = \frac{\mathrm{d}}{\mathrm{d}t} \int_{\mathbb{R}^3} W^{\mathrm{H}}(\mathbb{C}, \mathbf{H}, \boldsymbol{\xi}) \, \mathrm{d}\mathcal{V}_0 - \int_{\partial \mathcal{V}_0 \setminus \partial \mathcal{V}_0^{\mathbf{u}}} \mathbf{T} \cdot \dot{\mathbf{u}} \, \mathrm{d}\mathcal{S}_0. \tag{26}$$

Here, $W^{\mathrm{H}}(\mathbb{C}, \mathbf{H}, \boldsymbol{\xi})$ is the local potential energy density, where $\mathbb{C} = \mathbf{F}^T \mathbf{F}$ is the right Cauchy-Green tensor and $\mathbf{T}$ is the mechanical traction. The operators $(\dot{\blacksquare})$ and $\mathrm{d}/\mathrm{d}t$ in (26) denote the material time derivative. Notice that the local energy density $W^{\mathrm{H}}(\mathbf{F}, \mathbf{H}, \boldsymbol{\xi})$ is non-zero both inside and outside $\mathcal{V}_0$, which is typical in the magneto-mechanical formulation (Kankanala & Triantafyllidis, 2004). Nonetheless, for (26) to be bounded, the energy W^{H} must vanish as $|\mathbf{X}| \to \infty$, which is a realistic condition. In practice, in realistic and numerical boundary value problems, we always analyze finite domains that are subsets of $\mathbb{R}^3$. Those domains are sufficiently large allowing for the magnetic fields to reach an approximately zero value at their external boundary. More discussion about this point will be carried out further below.

The dissipation potential $\mathscr{D}$ associated with the solid is also given in terms of the local dissipation potential D, such that

$$\mathscr{D} = \int_{\mathcal{V}_0} D(\dot{\boldsymbol{\xi}}; \mathbb{C}, \mathbf{H}, \boldsymbol{\xi}) \, \mathrm{d}\mathcal{V}_0. \tag{27}$$

[2] For the moment, we keep $\boldsymbol{\xi}$ general and a more precise choice will be done further below. This choice depends on the dissipative mechanisms that are modeled. For instance viscous strains and/or magnetic dissipation.

Notice that the dissipation is only considered inside the solid domains, whereas it vanishes identically for all $\mathbf{X} \in \mathbb{R}^3 \backslash \mathcal{V}_0$. In addition, it is a principal function of the rate of the internal variable $\dot{\boldsymbol{\xi}}$, while the use of $\mathbb{C}$,$\mathbf{H}$ and $\boldsymbol{\xi}$ indicates mainly a history dependence.

With the above definitions in place, we propose an incremental variational principle following the seminal works of Onsager (1931a; 1931b), such that

$$\dot{\Pi}^{\mathrm{H}} = \inf_{\dot{\mathbf{u}} \in \widetilde{\mathcal{U}}} \sup_{\dot{\varphi} \in \widetilde{\mathcal{G}}} \inf_{\dot{\boldsymbol{\xi}}} \left[\dot{\mathscr{P}}^{\mathrm{H}} + \mathscr{D} \right]. \tag{28}$$

The admissible sets for $\dot{\mathbf{u}}$ and $\dot{\varphi}$ are given by

$$\widetilde{\mathcal{U}} \equiv \Big\{ \dot{\mathbf{u}}(\mathbf{X}) : \dot{\mathbf{F}}(\mathbf{X}) = \mathrm{Grad}\, \dot{\mathbf{u}}(\mathbf{X}) \;\; \forall\, \mathbf{X} \in \mathbb{R}^3, \quad \pi(\dot{\mathbf{u}}(\mathbf{X})) = \pi(\dot{\overline{\mathbf{u}}}(\mathbf{X})) \;\; \forall\, \mathbf{X} \in \partial\mathcal{V}_0^{\mathbf{u}} \Big\} \tag{29}$$

and

$$\widetilde{\mathcal{G}} \equiv \Big\{ \dot{\varphi}(\mathbf{X}) : \dot{\mathbf{H}}(\mathbf{X}) = -\mathrm{Grad}\, \dot{\varphi}(\mathbf{X}) \;\; \forall\, \mathbf{X} \in \mathbb{R}^3, \quad \dot{\varphi}(\mathbf{X}) = \dot{\overline{\varphi}}(\mathbf{X}) \;\; \forall\, \mathbf{X} \in \partial\mathcal{V}_\infty^{\varphi} \Big\}, \tag{30}$$

respectively. The symbol π in (29) denotes a projection operator that enables applying the constraints only on certain components on $\dot{\mathbf{u}}$ for all $\mathbf{X}$ on the domain boundary $\partial\mathcal{V}_0^{\mathbf{u}}$, where the displacement is constrained to vary with a reference rate $\dot{\overline{\mathbf{u}}}$. Similarly, the boundary where the magnetic potential φ varies according to a given rate $\dot{\overline{\varphi}}$ is denoted as $\partial\mathcal{V}_\infty^{\varphi}$. Since the magnetic fields are typically applied at a distance (and usually far away) from the MRE sample, we consider $\partial\mathcal{V}_\infty^{\varphi}$ to be a boundary surface enclosing the solid's boundary $\partial\mathcal{V}_0$. This particular consideration, of course, does *not* affect the generality of the variational principle (28) and different boundary conditions on φ may be imposed.

The stationarity conditions for $\dot{\Pi}^{\mathrm{H}}$ along with the arbitrariness of the considered volume element in $\mathcal{V}_0$ leads to the local governing equations and the boundary conditions in this scalar potential formulation. Thus, straightforward algebraic manipulations (see Kankanala and Triantafyllidis (2004) or Bustamante et al. (2008) for instance) leads to

$$\mathrm{Div}\, \mathbf{S} = \mathbf{0} \text{ in } \mathbb{R}^3, \quad \mathbf{S} = 2\mathbf{F}\frac{\partial W^{\mathrm{H}}}{\partial \mathbb{C}}, \quad [\![\mathbf{S}]\!] \cdot \boldsymbol{\mathcal{N}} - \mathbf{T} = \mathbf{0} \text{ on } \partial\mathcal{V}_0 \backslash \partial\mathcal{V}_0^{\mathbf{u}}, \tag{31}$$

$$\mathrm{Div}\, \mathbf{B} = 0 \text{ in } \mathbb{R}^3, \quad \mathbf{B} = -\frac{\partial W^{\mathrm{H}}}{\partial \mathbf{H}}, \quad [\![\mathbf{B}]\!] \cdot \boldsymbol{\mathcal{N}} = 0 \text{ on } \partial\mathcal{V}_0, \tag{32}$$

$$\frac{\partial W^{\mathrm{H}}}{\partial \boldsymbol{\xi}} + \frac{\partial D}{\partial \dot{\boldsymbol{\xi}}} = \mathbf{0} \quad \text{for all } \mathbf{X} \in \mathcal{V}_0, \quad \text{with } \boldsymbol{\mathcal{B}}^r = \frac{\partial D}{\partial \dot{\boldsymbol{\xi}}} = -\frac{\partial W^{\mathrm{H}}}{\partial \boldsymbol{\xi}}. \tag{33}$$

Here, $\mathcal{B}^r$ is the work conjugate to the corresponding internal variable $\boldsymbol{\xi}$. We see thus that the stationarity of $\dot{\Pi}^{\mathrm{H}}$ leads to the field balance equations and boundary conditions described in Sect. 2. In addition, it provides via the definition of the energy density W^{H}, the corresponding constitutive relations. Finally, the variational principle also leads to the GSM relation $(33)_1$, which provides, in fact, the local evolution equation for the internal variable $\boldsymbol{\xi}$. Notice that unlike the primary variables $\mathbf{u}$ and φ, the internal variable $\boldsymbol{\xi}$ *does not* need to satisfy any differential or boundary constraints. Moreover, the evolution Eq. $(33)_1$ only holds in the MRE domain, i.e., for all $\mathbf{X} \in \mathcal{V}_0$.

Entropy imbalance and constraint on $D(\dot{\boldsymbol{\xi}})$: The local form of the entropy imbalance equation, also known as the Clausius-Duhem inequality reads for the **F–H** model

$$\mathbf{S} : \dot{\mathbf{F}} - \mathbf{B} \cdot \dot{\mathbf{H}} - \dot{W}^{\mathrm{H}}(\mathbb{C}, \mathbf{H}, \boldsymbol{\xi}) \geq 0. \tag{34}$$

Expanding the derivative $\dot{W}^{\mathrm{H}}$ followed by substitutions of the constitutive relations $(31)_1$, $(32)_1$ and $(33)_1$ into (34), we obtain

$$\frac{\partial D}{\partial \dot{\boldsymbol{\xi}}} \cdot \dot{\boldsymbol{\xi}} \geq 0, \tag{35}$$

which is typically referred as the dissipation inequality. Notably, any function $D(\dot{\boldsymbol{\xi}}; \mathbb{C}, \mathbf{H}, \boldsymbol{\xi})$ that is convex in $\dot{\boldsymbol{\xi}}$ satisfies automatically the dissipation inequality, thus ensuring positive dissipation during any loading/unloading operation.

3.2 Vector Potential-Based F-B Formulation

In this section, we provide a dual formulation (under certain conditions) to the previous **F-H** one. Specifically, we derive the local balance laws and constitutive relations for an equivalent *dual* **F-B**-based formulation. Notice that, the **B** field is divergence-free and hence is now expressed in terms of a vector potential **A**, as defined in (24). In this formulation, we consider **u** and **A** to be the primary variables, whereas the internal variable remains the same, i.e., $\boldsymbol{\xi}$.

The rate of total potential energy is then given by

$$\dot{\mathscr{P}}^{\mathrm{B}} = \frac{\mathrm{d}}{\mathrm{d}t} \int_{\mathbb{R}^3} W^{\mathrm{B}}(\mathbb{C}, \mathbf{B}, \boldsymbol{\xi}) \, \mathrm{d}\mathcal{V}_0 - \int_{\partial \mathcal{V}_0} \mathbf{T} \cdot \dot{\mathbf{u}} \, \mathrm{d}\mathcal{S}_0 \tag{36}$$

with

$$W^{\mathrm{B}}(\mathbb{C}, \mathbf{B}, \boldsymbol{\xi}) = \sup_{\mathbf{H}} \Big[W^{\mathrm{H}}(\mathbb{C}, \mathbf{H}, \boldsymbol{\xi}) + \mathbf{H} \cdot \mathbf{B} \Big], \tag{37}$$

which is the partial Legendre-Fenchel transform of $W^{\mathrm{H}}(\mathbb{C}, \mathbf{H}, \boldsymbol{\xi})$ with respect to **H** (Bustamante et al., 2008). In turn, the dissipation potential, $\mathscr{D}$, is defined only in terms of the rate of the internal variable $\dot{\boldsymbol{\xi}}$ from Eq. (27) and thus is identical to that introduced in the **F-H** formulation. Thus, the minimization variational principle in terms of $\dot{\mathscr{P}}^{\mathrm{B}}$ and $\mathscr{D}$ now reads

$$\dot{\Pi}^{\mathrm{B}} = \inf_{\dot{\mathbf{u}}\in\widetilde{\mathcal{U}}} \inf_{\dot{\mathbf{A}}\in\widetilde{\mathcal{B}}} \inf_{\dot{\boldsymbol{\xi}}} \left[\dot{\mathscr{P}}^{\mathrm{B}} + \mathscr{D} \right], \tag{38}$$

where the admissible set $\widetilde{\mathcal{U}}$ for the rate $\dot{\mathbf{u}}$ is given by (29) and that for $\dot{\mathbf{A}}$ reads[3]

$$\widetilde{\mathcal{B}} \equiv \left\{ \dot{\mathbf{A}}(\mathbf{X}) : \dot{\mathbf{B}}(\mathbf{X}) = \mathrm{Curl}\, \dot{\mathbf{A}}(\mathbf{X}) \ \forall\, \mathbf{X} \in \mathbb{R}^3,\ \dot{\mathbf{A}}(\mathbf{X}) = \dot{\overline{\mathbf{A}}}(\mathbf{X}) \ \forall\, \mathbf{X} \in \partial\mathcal{V}_{\infty}^{\mathbf{A}} \right\}, \tag{39}$$

where the specific rate $\dot{\overline{\mathbf{A}}}$ is considered on the boundary $\partial\mathcal{V}_{\infty}^{\mathbf{A}}$ that encloses the MRE solid boundary $\partial\mathcal{V}_0$. Note further that contrary to (26) which is a saddle point stationary principle, the **F-B** potential (36) is a purely minimization principle.

In this regard, minimization of $\dot{\Pi}^{\mathrm{B}}$ with respect to the rates $\dot{\mathbf{u}}$, $\dot{\mathbf{A}}$ and $\dot{\boldsymbol{\xi}}$ leads to the local balance laws, constitutive relations along with the boundary conditions, so that

$$\mathrm{Div}\, \mathbf{S} = \mathbf{0} \text{ in } \mathbb{R}^3, \quad \mathbf{S} = 2\mathbf{F}\frac{\partial W^{\mathrm{B}}}{\partial \mathbb{C}}, \quad [[\mathbf{S}]] \cdot \boldsymbol{\mathcal{N}} - \mathbf{T} = \mathbf{0} \text{ on } \partial\mathcal{V}_0 \backslash \partial\mathcal{V}_0^{\mathbf{u}}, \tag{40}$$

$$\mathrm{Curl}\, \mathbf{H} = \mathbf{0} \text{ in } \mathbb{R}^3, \quad \mathbf{H} = \frac{\partial W^{\mathrm{B}}}{\partial \mathbf{B}}, \quad \boldsymbol{\mathcal{N}} \times [[\mathbf{H}]] = \mathbf{0} \text{ on } \partial\mathcal{V}_0, \tag{41}$$

$$\frac{\partial W^{\mathrm{B}}}{\partial \boldsymbol{\xi}} + \frac{\partial D}{\partial \dot{\boldsymbol{\xi}}} = \mathbf{0} \quad \text{for all } \mathbf{X} \in \mathcal{V}_0, \quad \text{with } \boldsymbol{\mathcal{B}}^r = \frac{\partial D}{\partial \dot{\boldsymbol{\xi}}} = -\frac{\partial W^{\mathrm{B}}}{\partial \boldsymbol{\xi}}. \tag{42}$$

Notice that (40) and (42) remain identical to (31) and (33), respectively, with the only difference being the replacement of W^{H} with W^{B}. The minimization of $\dot{\Pi}^{\mathrm{B}}$ with respect to $\dot{\mathbf{A}}$ yields the local balance law $(41)_1$, constitutive relation $(41)_2$ and the interface/boundary condition $(41)_3$ on $\partial\mathcal{V}_0$.

The **F-B** version of the local Clausius-Duhem inequality can be readily obtained by substituting (37) into (34). In turn, the dissipation inequality can be derived *mutatis mutandis* from the **F-H** case. In fact, the final form of the dissipation inequality remains identical to (35).

Remark 3.2 The above potential energy may be modified in a straighforward manner to include the Eddy current approximation and free volume currents $\mathbf{J}$, such that

$$\dot{\mathscr{P}}^{\mathrm{B}} = \frac{\mathrm{d}}{\mathrm{d}t} \int_{\mathbb{R}^3} W^{\mathrm{B}}(\mathbb{C}, \mathbf{B}, \boldsymbol{\xi})\, \mathrm{d}\mathcal{V}_0 - \int_{\partial\mathcal{V}_0} \mathbf{T} \cdot \dot{\mathbf{u}}\, \mathrm{d}\mathcal{S}_0 - \int_{\mathcal{V}_0} \mathbf{J} \cdot \dot{\mathbf{A}}\, \mathrm{d}\mathcal{V}_0. \tag{43}$$

[3] The numerical solution for the three-dimensional vector potential-based BVPs requires an additional constraint on $\mathbf{A}$, namely, $\mathrm{Div}\, \mathbf{A} = 0$, commonly referred as the Coulomb gauge. The latter is not necessary in a two-dimensional problem, whereas the three dimensional implementation is discussed further below.

Then the only equation changing in the local balance laws is $(41)_1$, which becomes $\text{Curl}\,\mathbf{H} = \mathbf{J}$ in $\mathcal{V}_0$. The free volume current is effectively a body force term in domains that the magnetic vector potential $\dot{\mathbf{A}} \neq \mathbf{0}$.

4 Modeling of Isotropic Hard-MREs

In this section, we lay out the constitutive laws necessary to describe the response of isotropic incompressible h-MREs. The procedure is of course general, however, the choice of functions pertains *only* to the specific class of materials analyzed here.

Before we proceed to specific constitutive propositions, we discuss first the constraints that need to be imposed on the energetic and dissipation potentials, W^{H} or W^{B} and D, respectively, in order to ensure (a) an even magneto-mechanical coupling, (b) material frame indifference, (c) isotropic material symmetry and (d) positive dissipation. For conciseness, we will use the symbol

$$\mathbf{G} = \begin{cases} \mathbf{H}, & \text{if } \mathbf{F}\text{–}\mathbf{H} \text{ formulation} \\ \mathbf{B}, & \text{if } \mathbf{F}\text{–}\mathbf{B} \text{ formulation} \end{cases} \tag{44}$$

to denote compactly the two formulations, wherever that is possible.

4.1 Internal Variable for Magnetic Dissipation

A thermodynamically consistent model for any dissipative material may be constructed through the definition of a finite number of internal variables, which reflect the irreversible processes the material undergoes under external loads. In this regard, one of the principal differences between h-MREs and s-MREs is the underlying magnetic dissipation of the filler particles (e.g. NdFeB) in the former. Due to the finite strains and the magneto-mechanical coupling, upon cyclic magnetic loading, the response of the h-MRE composite exhibits both magnetic and mechanical (due to magneto-mechanical coupling) hysteresis.

In the work of Mukherjee et al. (2021), it was shown via extensive assessment with numerical RVE simulations that only one internal *magnetic* vectorial variable suffices to describe both the magnetic and the mechanical (due to coupling) dissipation in the h-MRE in the case of moderately soft to hard polymer matrices. This is a mere *constitutive choice* and does not constitute a unique way to address this coupled dissipative effect.[4]

[4] Works by McMeeking and Landis (2005), McMeeking et al. (2007), Rosato and Miehe (2014) in the context of ferroelectricity, a somewhat similar problem, use both mechanical and polarization internal variables that are eventually related via additional constitutive relations.

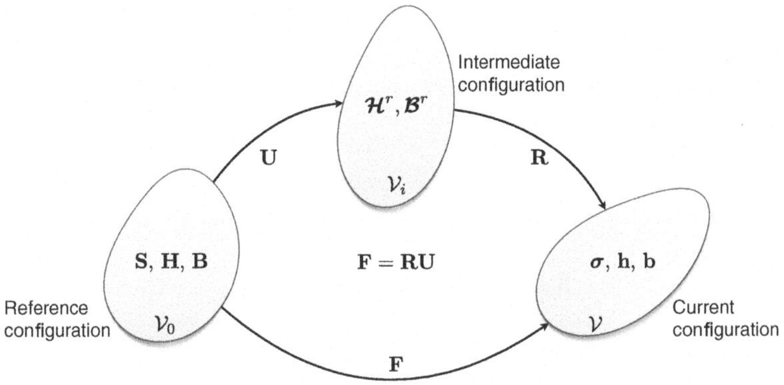

Fig. 2 Definition of the reference, intermediate and current configurations of volume $\mathcal{V}_0$, $\mathcal{V}_i$ and $\mathcal{V}$, respectively, along with the different field variables defined therein

Specifically, we consider that the internal variable is a vector,

$$\boldsymbol{\xi} \equiv \boldsymbol{\mathcal{H}}^r \in \mathbb{R}^3, \tag{45}$$

which lies in the *stretch-free*, *intermediate* configuration $\mathcal{V}_i$, as shown in Fig. 2. This internal variable is a remanent H-like vectorial quantity that will be shown in practice to be strongly linked to the magnetization of the MRE. The central assumption behind this "choice" is that the average magnetization of the composite and thus magnetic dissipation is affected by macroscopic rotations of the magnetization but is independent of macroscopic stretches in the specific case of *incompressible* MREs. A more detailed discussion of the consequences of this assumption will be carried out in Sect. 7 in parallel with the presentation of specific results.

The range of validity of this assumption depends mainly on the shear modulus of the matrix phase, given that the particles are almost mechanically rigid and thus do not deform but may rotate, whereas the bulk modulus of the matrix phase is very large giving rise to a quasi-incompressible macroscopic response of the h-MRE. In fact, the shear modulus of the matrix controls the capability of the magnetized particles to rotate upon application of a magnetic load that is non-aligned with their magnetization vector. In a soft polymeric matrix, the particles rotate mechanically but also evolve their magnetic state via dissipative mechanisms (such as domain wall motions) in order to align their magnetization with the externally applied magnetic field (Kalina et al., 2017). The softer the matrix the more the rotations prevail over the magnetic dissipation mechanisms. By contrast, when the shear modulus of the matrix has a moderate value (e.g. greater than $\sim$100 kPa), local particle rotations are much less pronounced (provided that the particles are not very elongated either) and dissipative processes dominate the response of the h-MRE, while any rotations of the magnetized particles follow approximately the overall macroscopic rotation. This last case corresponds well to actual, fabricated h-MREs in the literature (see for instance Zhao et al. (2019), Ren et al. (2019), Garcia-Gonzalez et al. (2023)).

In view of this discussion, use of the standard polar decomposition $\mathbf{F} = \mathbf{RU}$ directly implies that the reference (in $\mathcal{V}_0$) and current (in $\mathcal{V}$) remanent fields, $\mathbf{H}^r$ and $\mathbf{h}^r$, respectively are functions of $\mathcal{H}^r$ and are given by

$$\mathbf{h}^r = \mathbf{R}\mathcal{H}^r, \qquad \mathbf{H}^r = \mathbf{U}\mathcal{H}^r. \tag{46}$$

whereby the following standard decompositions also hold (Mukherjee & Danas, 2019)

$$\mathbf{h} = \mathbf{h}^e + \mathbf{h}^r, \qquad \mathbf{H} = \mathbf{H}^e + \mathbf{H}^r. \tag{47}$$

Here, $\mathbf{h}$ and $\mathbf{h}^e$ are the Eulerian total and energetic h-fields, while $\mathbf{H}$ and $\mathbf{H}^e$ are the corresponding Lagrangian ones. The introduction of $\mathcal{H}^r$ in the *intermediate* configuration $\mathcal{V}_i$ further implies that the Eulerian $\mathbf{h}^r$ is stretch-free function of $\mathbf{R}$ and $\mathcal{H}^r$, while the Lagrangian $\mathbf{H}^r$ is a function of $\mathbf{U}$ and $\mathcal{H}^r$.

Obviously, the propositions (46) remain open to further refinements if required by the problem at hand. Nonetheless, we will show in the results presented in Sect. 7 that the single internal variable $\mathcal{H}^r$ is sufficient to achieve very good agreement with the corresponding numerical periodic cell simulations under a wide range of cyclic and non-aligned loading conditions. Addition of compressibility in the matrix is expected to change this feature and in that case extensions of the aforementioned choices shall be necessary. Finally, with this observation, the standard materials relations (33) and (42) are also defined in $\mathcal{V}_i$ instead of $\mathcal{V}_0$. In this view, the conjugate variable to $\mathcal{H}^r$ will be denoted for consistency with $\mathcal{B}^r$ and is fairly similar to the backstress in purely mechanical elasto-plastic systems. The relevant GSM relations (33) and and (42) then become

$$\frac{\partial W^{\mathrm{G}}}{\partial \mathcal{H}^r} + \frac{\partial D}{\partial \dot{\mathcal{H}}^r} = \mathbf{0} \text{ for all } \mathbf{X} \in \mathcal{V}_0, \quad \text{with } \mathcal{B}^r = \frac{\partial D}{\partial \dot{\mathcal{H}}^r} = -\frac{\partial W^{\mathrm{G}}}{\partial \mathcal{H}^r}, \tag{48}$$

where again $\mathbf{G} = \{\mathbf{H}, \mathbf{B}\}$ has been introduced in (44).

Remark 4.1 The recent work of Mukherjee and Danas (2022) denotes the internal variable by $\boldsymbol{\xi} \equiv \mathcal{H}^r$. Given the fact that the dissipation potential D is identical in both the **F**-**H** and the **F**-**B** formulations, a choice of a neutral notation was done there to avoid confusion. Nonetheless, we maintain in this manuscript the original $\mathcal{H}^r$ one, since in future works $\boldsymbol{\xi}$ may be used to denote additional internal variables as is the case of viscoelasticity (Rambausek et al., 2022; Lucarini et al., 2022a).

4.2 General Properties of the Free Energy Density and the Dissipation Potential

We recall first that material objectivity and symmetry conditions are well-known for s-MREs (Kankanala & Triantafyllidis, 2004; Dorfmann & Ogden, 2004). However, as shown in Fig. 2, the present models for h-MREs introduce a remanent internal variable that is insensitive to the stretch $\mathbf{U}$, leading to non-familiar magneto-mechanical invariants and rates. Hence, this section focuses on stating explicitly all the aforementioned constraints in the context of isotropic h-MREs. The proposed invariants may then be simplified to attain those that remain also valid for s-MREs.

Even magneto-mechanical coupling. The magneto-mechanical energy density W^{G} and dissipation potential D must be exactly the same when both $\mathbf{G}$ and $\mathcal{H}^r$ change simultaneously sign. This condition reads

$$W^{\mathrm{G}}(\mathbb{C}, -\mathbf{G}, -\mathcal{H}^r) = W^{\mathrm{G}}(\mathbb{C}, \mathbf{G}, \mathcal{H}^r), \qquad D(-\dot{\mathcal{H}}^r) = D(\dot{\mathcal{H}}^r), \tag{49}$$

and ensures a symmetric, butterfly-type magnetostriction response for the h-MRE, as will be discussed in the results Sect. 7.

Material frame indifference. This physical property imposes that W and D must remain invariant under a change of observer. A change of observer leads to the new current position vector $\mathbf{x}^* = \mathbf{c} + \mathbf{Qx}$, where $\mathbf{c}$ is a rigid displacement field and $\mathbf{Q}$ is a proper rotation matrix (Gurtin 1982, pp. 139–142). Since the arguments of W^{G} and D, i.e. $\mathbb{C}$, $\mathbf{G}$, $\mathcal{H}^r$ and $\dot{\mathcal{H}}^r$ are unaffected by such a transform, the requirement of material frame indifference imposes no further restrictions on W^{G} and D. This observation is in agreement with the objectivity conditions used in mechanical visco-plasticity, where the intermediate strain-like variables remain unaffected by a change of observer (Dashner, 1993; Kumar & Lopez-Pamies, 2016).

Material symmetry. For isotropic MREs, W^{G} and D must remain invariant under a change in the reference configuration via a constant matrix $\mathbf{K} \in Orth^+$. The material symmetry conditions on the potentials thus read

$$\begin{aligned} &W^{\mathrm{G}}(\mathbf{K}^T\mathbb{C}\mathbf{K}, \mathbf{K}^T\mathbf{G}, \mathbf{K}^T\mathcal{H}^r) = W(\mathbb{C}, \mathbf{G}, \mathcal{H}^r), \\ &D(\mathbf{K}^T\dot{\mathcal{H}}^r; \mathbf{K}^T\mathbb{C}\mathbf{K}, \mathbf{K}^T\mathbf{G}, \mathbf{K}^T\mathcal{H}^r) = D(\dot{\mathcal{H}}^r; \mathbb{C}, \mathbf{G}, \mathcal{H}^r). \end{aligned} \tag{50}$$

Here, we note that the intermediate $\mathcal{H}^r$ transforms as $\mathcal{H}^r \to \mathbf{K}^T\mathcal{H}^r$, which follows from (46) and the transformation $\mathbf{U} \to \mathbf{K}^T\mathbf{U}\mathbf{K}$ under a change in the reference configuration. We further remark that in mechanical visco-plasticity the change in the reference configuration also modifies the intermediate plastic internal variables (Dashner, 1993; Bennett et al., 2016).

4.3 The Isotropic Magneto-Mechanical Invariants for h-MREs

A straightforward and elegant way to satisfy the conditions of even magneto-mechanical coupling, isotropic material symmetry and frame indifference is to express the energy density and dissipation in terms of properly chosen *isotropic* invariants. First, we define the general set of available invariants given the corresponding arguments $\mathbb{C}$, $\mathbf{G}$ and $\mathcal{H}^r$. Subsequently, a reduced number of them will be chosen to model the h-MREs. The choice is mainly motivated by corresponding numerical RVE simulations of two-phase h-MRE composites and does not constitute a rigorous result but merely an efficient homogenization-guided strategy that allows to maintain the number of invariants small and the model entirely explicit.

Mechanical invariants.

$$I_1 = \mathrm{tr}(\mathbb{C}), \qquad I_2 = \mathrm{tr}(\mathbb{C})^2 - \mathrm{tr}(\mathbb{C}^2), \qquad I_3 = J^2 = \det \mathbb{C} = 1, \tag{51}$$

Magneto-mechanical invariants in **F**-**H** *formulation.*

$$\begin{aligned} I_4^{\mathrm{H}} &= \mathbf{H}\cdot\mathbf{H}, & I_4^{\mathrm{HHr}} &= \mathbf{H}\cdot\mathbb{C}^{1/2}\mathcal{H}^r, & I_4^{\mathrm{Hr}} &= \mathcal{H}^r\cdot\mathbb{C}\mathcal{H}^r \\ I_5^{\mathrm{H}} &= \mathbf{H}\cdot\mathbb{C}^{-1}\mathbf{H}, & I_5^{\mathrm{HHr}} &= \mathbf{H}\cdot\mathbb{C}^{-1/2}\mathcal{H}^r, & I_5^{\mathrm{Hr}} &= \mathcal{H}^r\cdot\mathcal{H}^r. \\ I_6^{\mathrm{H}} &= \mathbf{H}\cdot\mathbb{C}\mathbf{H}, & I_6^{\mathrm{HHr}} &= \mathbf{H}\cdot\mathbb{C}^{3/2}\mathcal{H}^r, & I_6^{\mathrm{Hr}} &= \mathcal{H}^r\cdot\mathbb{C}^2\mathcal{H}^r. \end{aligned} \tag{52}$$

Magneto-mechanical invariants in **F**-**B** *formulation.*

$$\begin{aligned} I_4^{\mathrm{B}} &= \mathbf{B}\cdot\mathbf{B}, & I_4^{\mathrm{BHr}} &= \mathbf{B}\cdot\mathbb{C}^{-1/2}\mathcal{H}^r, & I_4^{\mathrm{Hr}} &= \mathcal{H}^r\cdot\mathbb{C}\mathcal{H}^r \\ I_5^{\mathrm{B}} &= \mathbf{B}\cdot\mathbb{C}\mathbf{B}, & I_5^{\mathrm{BHr}} &= \mathbf{B}\cdot\mathbb{C}^{1/2}\mathcal{H}^r, & I_5^{\mathrm{Hr}} &= \mathcal{H}^r\cdot\mathcal{H}^r. \\ I_6^{\mathrm{B}} &= \mathbf{B}\cdot\mathbb{C}^2\mathbf{B}, & I_6^{\mathrm{BHr}} &= \mathbf{B}\cdot\mathbb{C}^{3/2}\mathcal{H}^r & I_6^{\mathrm{Hr}} &= \mathcal{H}^r\cdot\mathbb{C}^2\mathcal{H}^r. \end{aligned} \tag{53}$$

The energetic invariants, i.e., those that do not involve $\mathcal{H}^r$, are the standard ones employed usually in the context of s-MREs (Kankanala & Triantafyllidis, 2004; Ponte Castañeda & Galipeau, 2011; Danas et al., 2012; Danas, 2017; Mukherjee et al., 2020). The remaining invariants are mixed or purely remanent ones, and are necessary in the modeling of h-MREs. We note here that the use of the rational exponents in the mixed invariants are chosen in order to allow the proper amplitude of coupling. Moreover, it is emphasized that $\mathcal{H}^r$ has the same units as the h-field and thus proper addition of μ_0 is required in the final energy expressions. Finally, in order to satisfy the invariance of the dissipation potential, we additionally employ the Euclidean norm $|\dot{\mathcal{H}}^r| = \sqrt{\dot{\mathcal{H}}^r\cdot\dot{\mathcal{H}}^r}$ as an invariant of $\dot{\mathcal{H}}^r$.

We further note that the "uncoupled" I_5-type invariants are unaffected by mechanical deformation when a certain Eulerian magnetic field is independently controlled whereas the I_4 do vary in this case (see relevant discussion in Danas (2017)).

4.4 Form of Energy Densities

We now express the energy densities associated with an incompressible h-MRE in terms of these invariants. In particular, both W^{H} and W^{B} are considered to be the sum of three distinct energy densities, namely, the purely mechanical, purely magnetic and coupling free energy density, such that

$$W^{\mathrm{H}}(\mathbb{C}, \mathbf{H}, \mathcal{H}^r) = \rho_0 \Psi_{\mathtt{mech}}(I_1) + \rho_0 \Psi^{\mathrm{H}}_{\mathtt{mag}}(I_5^{\mathrm{H}}, I_5^{\mathrm{HHr}}, I_5^{\mathrm{Hr}}) + \\ + \rho_0 \Psi^{\mathrm{H}}_{\mathtt{couple}}(I_4^{\mathrm{HHr}}, I_4^{\mathrm{Hr}}, I_5^{\mathrm{HHr}}, I_5^{\mathrm{Hr}}) - \frac{\mu_0}{2} I_5^{\mathrm{H}} \tag{54}$$

and

$$W^{\mathrm{B}}(\mathbb{C}, \mathbf{B}, \mathcal{H}^r) = \rho_0 \Psi_{\mathtt{mech}}(I_1) + \rho_0 \Psi^{\mathrm{B}}_{\mathtt{mag}}(I_5^{\mathrm{B}}, I_5^{\mathrm{BHr}}, I_5^{\mathrm{Hr}}) \\ + \rho_0 \Psi^{\mathrm{B}}_{\mathtt{couple}}(I_4^{\mathrm{Hr}}, I_5^{\mathrm{BHr}}, I_5^{\mathrm{Hr}}, I_6^{\mathrm{BHr}}, I_6^{\mathrm{Hr}}) + \frac{1}{2\mu_0} I_5^{\mathrm{B}}. \tag{55}$$

We will show later in this section that W^{B} is an exact dual of W^{H} in the sense defined in $(36)_2$. For this to be possible, one should have fairly simple expressions that allow for an analytical Legendre-Fenchel transform. That will be the case here but it is not always possible, as we will show in the case of purely energetic s-MREs in Sect. 5.

In both expressions, ρ_0 is the reference density of the solid, while the last terms $\mu_0 I_5^{\mathrm{H}}/2$ in (54) and $I_5^{\mathrm{B}}/2\mu_0$ in (55) represent the energy associated with free space with μ_0 being the magnetic permeability in vacuum or in non-magnetic solids such as the polymer matrix phase.

Remark 4.2 In absence of any magnetic material, $\Psi^{\mathrm{G}}_{\mathtt{mag}} = \Psi^{\mathrm{G}}_{\mathtt{couple}} = 0$, however the energy density W^{H} or W^{B} does not vanish. Instead the presence of the last term in (54) or (55) allows the magnetic fields to exist in the vacuum or in non-magnetic materials, such as the matrix phase. These terms are in general necessary even if one works with a magnetization based formulation. In formulations that these terms are omitted, the effect of the surrounding space or neighboring non-magnetic materials are in general not taken into account. This, however, is primordial in the present case of the modeling of particle-filled MREs, since these terms allow to retain the long-range interactions between the particles. Such interactions in nonlinear materials are in general non-trivial and cannot be fully described by simpler particle-particle (nearest neighbors) dipole-bases models.

The introduction of such an incompressible material model necessitates the modification of the constitutive relations $(31)_2$ and $(40)_2$ for the total stress $\mathbf{S}$, such that

$$\mathbf{S} = 2\mathbf{F}\frac{\partial W^{\mathrm{G}}}{\partial \mathbb{C}} + p\mathbf{F}^{-T}, \tag{56}$$

where the superscript G refers the notation introduced in (44) and p is the Lagrange multiplier associated with the incompressibility constraint $J = 1$. In practice, p adds on to the local (point wise) number of unknowns to be solved after optimization of the variational principle.

Remark 4.3 We note that the *quasi-incompressible* equivalents of the proposed incompressible models are often useful in the numerical computations. Thus, we extend the proposed **F**-**H** and **F**-**B**-based MRE models in a rather ad-hoc way so that the energy densities read

$$W^{\text{H}}_{\text{comp}}(\mathbb{C}, \mathbf{H}, \mathcal{H}^r) = \rho_0 \Psi^{\text{comp}}_{\text{mech}}(I_1, J) + \rho_0 \Psi^{\text{H}}_{\text{mag}}(I_5^{\text{H}}, I_5^{\text{HHr}}, I_5^{\text{Hr}}) + \rho_0 \Psi^{\text{H}}_{\text{couple}}(I_4^{\text{HHr}}, I_4^{\text{Hr}}, I_5^{\text{HHr}}, I_5^{\text{Hr}}) - \frac{\mu_0}{2} J I_5^{\text{H}} \tag{57}$$

and

$$W^{\text{B}}_{\text{comp}}(\mathbb{C}, \mathbf{H}, \mathcal{H}^r) = \rho_0 \Psi^{\text{comp}}_{\text{mech}}(I_1, J) + \rho_0 \Psi^{\text{B}}_{\text{mag}}(I_5^{\text{B}}, I_5^{\text{BHr}}, I_5^{\text{Hr}}) + \rho_0 \Psi^{\text{B}}_{\text{couple}}(I_4^{\text{Hr}}, I_5^{\text{BHr}}, I_5^{\text{Hr}}, I_6^{\text{BHr}}) + \frac{I_5^{\text{B}}}{2\mu_0 J}, \tag{58}$$

respectively. Note that the proposed quasi-incompressible energy functions *are not meant to be used* for compressible MREs, since the proposed expressions do not take into account any coupling between volumetric deformations and magnetic fields. Such coupling requires an entirely independent analysis (Gebhart, P., & Wallmersperger, 2022a; 2022b) and more importantly should reflect a realizable compressible MRE. Currently, most MREs existing in literature are fairly incompressible except earlier and more recent studies on MRE foams (Diguet et al., 2021). Nonetheless, in those cases, taking into account the precise microstructure is of essence. Such studies are in progress and will be presented elsewhere in the future.

4.5 The Mechanical Energy Density

The purely mechanical free energy density $\rho_0 \Psi_{\text{mech}}$ is the same for both formulations and corresponds to the analytical homogenization estimate by Lopez-Pamies et al. (2013) for a two-phase composite made of an incompressible nonlinear elastic matrix comprising isotropic distributions of rigid-particles, such that

$$\Psi_{\text{mech}}(I_1; c) = (1-c)\Psi_{\text{m,mech}}(\mathcal{I}_1), \qquad \mathcal{I}_1 = \frac{I_1 - 3}{(1-c)^{7/2}} + 3, \tag{59}$$

where c is the particle volume fraction and $\Psi_{\text{m,mech}}$ is the free energy density of the matrix. Notably, the homogenization estimate (59) holds for any I_1-based incompressible rigid-particle–matrix composite. Thus, the choice for the matrix constitutive law remains versatile in the present modeling framework. Evidently, in the limit of $c = 0$, the homogenized energy recovers that of the matrix phase, i.e., $\Psi_{\texttt{mech}}(I_1, 0) = \Psi_{\texttt{m,mech}}(I_1)$. By contrast, $\Psi_{\texttt{mech}}(I_1, 1) = +\infty$ when $c = 1$, hence, recovering the energy of a rigid material. This part may be replaced readily by any other mechanical estimate, homogenization-based or phenomenological, that may be required.

An ad-hoc modification of the above incompressible energy allows to obtain a quasi-incompressible counterpart that proves useful in numerical computations. In particular, the term $\Psi^{\texttt{comp}}_{\texttt{mech}}(I_1, J)$ in (57) and (58), may be written as

$$\rho_0 \Psi^{\texttt{comp}}_{\texttt{mech}}(I_1, J) = (1-c)\rho_0 \Psi_{\texttt{m,mech}}(\mathcal{I}_1^{\texttt{comp}}) + \frac{G'_{\text{m}}}{2(1-c)^6}(J-1)^2,$$

$$\text{with} \quad \mathcal{I}_1^{\texttt{comp}} = \frac{I_1 - 3 - 2\ln J}{(1-c)^{7/2}} + 3. \tag{60}$$

In this expression, we typically set the Lamé parameter $G'_{\text{m}} \geq 200 G_{\text{m}}$, which ensures a quasi-incompressible response with $J \approx 1$.

4.6 *The Magnetic and Coupled Energy Densities*

Next, the magnetic and coupling free energy functions along with the dissipation potential are proposed. Special care is taken in proposing these functions so that the magnetization response yields several limiting cases, especially, in the limit of small primary and remanent magnetic fields. In the following, we provide these functions without elaborating on their individual significance.

4.6.1 F-H Expressions for *h*-MREs

The pure magnetic free energy is expressed in terms of the I_5-based invariants, so that

$$\rho_0 \Psi^{\text{H}}_{\texttt{mag}}(I_5^{\text{H}}, I_5^{\text{HHr}}, I_5^{\text{Hr}}) = -\frac{\mu_0}{2}\chi^e I_5^{\text{H}} + \mu_0(1+\chi^e) I_5^{\text{HHr}} + \frac{\mu_0}{2}\left(\frac{1-c}{3c}\right) I_5^{\text{Hr}}$$
$$+ \frac{\mu_0}{c}\frac{(m^s)^2}{\chi^r_{\text{p}}} f_{\text{p}}\left(\frac{\sqrt{I_5^{\text{Hr}}}}{m^s}\right). \tag{61}$$

Here χ^r_{p} is the *remanent susceptibility* of the underlying magnetic particle, whereas the "effective" parameters χ^e and m^s for the composite are given in terms of the particle magnetic properties and its volume fraction c as

$$\chi^e = \frac{3c\chi_p^e}{3+(1-c)\chi_p^e}, \qquad m^s = c\, m_p^s\left(\frac{1+\chi_p^e}{1+\chi^e}\right), \tag{62}$$

where χ_p^e and m_p^s are the particle *energetic susceptibility* and *saturation magnetization* (a graphic explanation of these parameters is provided later in the context of Fig. 3). Moreover, $f_p(x)$ is a nonlinear function that leads to a saturation-type magnetization behavior. Additionally, $f_p(x)$ must satisfy the properties such that (i) it is smooth and at least twice differentiable for all $0 \leq x < 1$, (ii) $f_p'(x)$ leading to an inverse saturation (sigmoid) function that tends to $+\infty$ in the limit of $x \to 1$ and (iii) the leading order Taylor series expansion of $f_p(x)$ around $x = 0$ goes as x^2. Of course, the specific choices for $f_p(x)$ depends on the saturation response of the (hard/soft) magnetic particles. A set of representative choices for such hard magnetic particles are provided in Table 1.

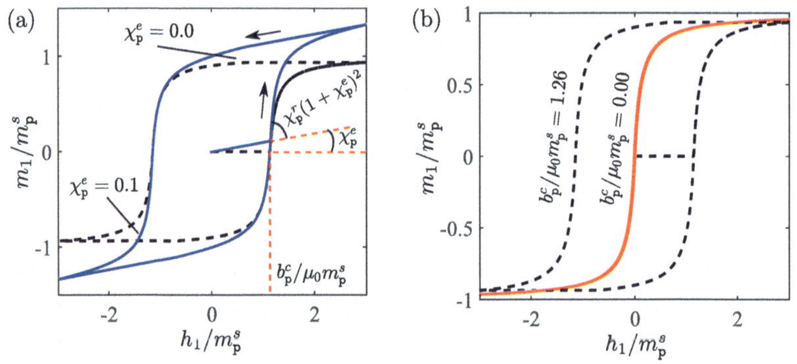

Fig. 3 **a** Magnetization response under applied uniaxial cyclic h-field $\mathbf{h} = h_1\mathbf{e}_1$. Both, ideal ($\chi_p^e = 0$) and actual ($\chi_p^e > 0$) hysteresis loops are shown along with the slopes of the $m - h$ response before and after switching. **b** Magnetization responses for finite and zero coercivity leading to, respectively, hysteretic and energetic magnetization responses. (Taken partly from Mukherjee and Danas (2022))

Table 1 Inverse saturation function choices for the remanent potential

Function name	$f_p(x)$
Inverse hypergeometric function	$-[\log(1-x)+x]$
Tangent function	$-\frac{4}{\pi^2}\log\left[\cos\left(\frac{\pi}{2}x\right)\right]$
Arctanh function	$-\left[(1-x)\tanh^{-1}(x) - \log(x+1)\right]$
Inverse Langevin function (approx.)	$\frac{1}{3}\Big(-\frac{x^3}{3} + x^2 - (0.0571x^2 - 0.0093)\cos(3.5x)$ $-x - \log(1-x) + 0.0327x\sin(3.5x) - 0.0093\Big)$

Finally, the coupling free energy $\Psi^{\mathrm{H}}_{\texttt{couple}}$ is proposed in terms of both, the I_4 and I_5-based invariants as defined in (52), so that

$$\rho_0 \Psi^{\mathrm{H}}_{\texttt{couple}}(I_4^{\mathrm{HHr}}, I_4^{\mathrm{Hr}}, I_5^{\mathrm{HHr}}, I_5^{\mathrm{Hr}}) = c\, \beta(c) \mu_0 \Big[\big(I_4^{\mathrm{Hr}} - I_5^{\mathrm{Hr}}\big) - 2\chi^e \big(I_4^{\mathrm{HHr}} - I_5^{\mathrm{HHr}}\big) \Big] \quad (63)$$

with

$$\beta(c) = 19.0c^2 - 10.4c + 1.71. \quad (64)$$

Notice that the I_4-type "coupling" invariants only appear in the coupling free energy $\Psi^{\mathrm{H}}_{\texttt{couple}}$, whereas, the purely magnetic free energy (61) is only a function of the "pure magnetic" or "decoupled" I_5-type invariants. Moreover, the coupling parameter $\beta(c)$ in (63) may be further calibrated against experimental data or numerical homogenization estimates. In this study, we will use the expression (64) as obtained by direct calibration with corresponding FE simulations in Sect. 7.1, and thus it is valid for $c \le 0.3$. We also note the simple linear dependence on the invariants of the coupled energy density in (63). This will prove extremely useful in obtaining a dual energy density for the **F-B** model in the following.

4.6.2 F-B Expressions for *h*-MREs

The proposed W^{H} in (54) along with the invariants in (52) lead to a *strictly concave* energy density function in terms of **H**, i.e., for a given $\mathbb{C}$ and $\mathcal{H}^r$, W^{H} is a strictly concave function of **H**. Note that the remaining invariants are functions of the internal variable $\mathcal{H}^r$, which remain exactly the same for the dual **F-B** version we describe here. Consequently, in order to obtain the equivalent, dual **F-B** energy density, we seek for a closed form partial Legendre-Fenchel transform of W^{H} following $(36)_2$. Straightforward algebraic manipulations lead to the expression for W^{B} in (55).

More specifically, the mechanical free energy $\Psi_{\texttt{mech}}$ in (55) remains identical to (59), whereas the transformed magnetic free energy $\Psi^{\mathrm{B}}_{\texttt{mag}}(I_5^{\mathrm{B}}, I_5^{\mathrm{BHr}}, I_5^{\mathrm{Hr}})$ becomes

$$\rho_0 \Psi^{\mathrm{B}}_{\texttt{mag}}(I_5^{\mathrm{B}}, I_5^{\mathrm{BHr}}, I_5^{\mathrm{Hr}}) = -\frac{1}{2\mu_0}\frac{\chi^e}{1+\chi^e} I_5^{\mathrm{B}} + I_5^{\mathrm{BHr}} + \frac{\mu_0}{2}\left(\chi^e + \frac{1+2c}{3c}\right) I_5^{\mathrm{Hr}} + \frac{\mu_0}{c}\frac{(m^s)^2}{\chi^r_{\mathrm{p}}} f_{\mathrm{p}}\left(\frac{\sqrt{I_5^{\mathrm{Hr}}}}{m^s}\right), \quad (65)$$

where all the model parameters along with the function $f_{\mathrm{p}}(x)$ remain identical to their respective definitions for the **F-H** model presented earlier. Finally, the coupling free energy reads

$$\rho_0 \Psi^{\mathrm{B}}_{\texttt{couple}}(I_4^{\mathrm{Hr}}, I_5^{\mathrm{BHr}}, I_5^{\mathrm{Hr}}, I_6^{\mathrm{BHr}}, I_6^{\mathrm{Hr}}) = c\beta\mu_0(1-2\chi^e)\Big\{I_4^{\mathrm{Hr}} - I_5^{\mathrm{Hr}}\Big\} - \frac{2c\beta\chi^e}{1+\chi^e}\Big\{I_6^{\mathrm{BHr}} - I_5^{\mathrm{BHr}}\Big\} + 2\mu_0\frac{(c\beta\chi^e)^2}{1+\chi^e}\Big\{I_5^{\mathrm{Hr}} + I_6^{\mathrm{Hr}} - 2I_4^{\mathrm{Hr}}\Big\}. \quad (66)$$

Notice further that the particle volume fraction remains limited to $c \leq 0.3$ in all practical applications of h-MREs (Zhao et al., 2019; Alapan et al., 2020). Consequently, we obtain $0 < \chi^e \ll 1$ for the h-MREs comprising NdFeB particles. Thus, the last term in the expression for $\rho_0 \Psi^{\mathrm{B}}_{\texttt{couple}}$ in (66) turns out to be substantially smaller than the preceding ones and hence, can be dropped for all practical modeling purposes. Consequently, the coupling energy in the proposed **F-B** model simplifies to

$$\rho_0 \Psi^{\mathrm{B}}_{\texttt{couple}}(I_4^{\mathrm{Hr}}, I_5^{\mathrm{BHr}}, I_5^{\mathrm{Hr}}, I_6^{\mathrm{BHr}}) = c\beta\left[\mu_0(1-2\chi^e)\left\{I_4^{\mathrm{Hr}} - I_5^{\mathrm{Hr}}\right\} - \frac{2\chi^e}{1+\chi^e}\left\{I_6^{\mathrm{BHr}} - I_5^{\mathrm{BHr}}\right\}\right]. \tag{67}$$

With this, the definition of the **F-B** based energy density becomes complete. Being the closed-form Legendre-Fenchel transform of the **F-H** model, the derived **F-B** model exhibits the *exact* same features as the corresponding **F-H** model discussed previously. Consequently, no further calibration of $\beta(c)$ parameter, defined in (64), is needed in (67). Specific comparisons between the local responses of the **F-H** and **F-B** models will be shown later in Sect. 7.1.

4.7 The Dissipation Potential

It remains to define the dissipation potential D, which along with W^{H} and W^{B} completes the model constitutive relations. Given that we do not include viscoelastic effects in the present manuscript, we propose the rate-dependent dissipation potential to be a simple power law in terms of $\dot{\mathcal{H}}^r$ only, such that (Mukherjee & Danas, 2019)

$$D(\dot{\mathcal{H}}^r) = \frac{n\, b^c}{n+1} |\dot{\mathcal{H}}^r|^{\frac{n+1}{n}}, \qquad \text{with } 1 \leq n < +\infty. \tag{68}$$

Here, $|.|$ is the standard Eulerian norm and b^c is the effective *coercive field* of the composite (see graphical representation in Fig. 3) that is given in terms of the particle and effective energetic susceptibility via

$$b^c = b^c_{\mathrm{p}} \left(\frac{1+\chi^e}{1+\chi^e_{\mathrm{p}}}\right)^{4/5}. \tag{69}$$

Here, b^c_{p} is the particle ceorcivity. Typically, for a hard-magnetic composite the effective coercivity is given by $b^c = b^c_{\mathrm{p}}$ (Idiart et al., 2006b). Nonetheless, the term multiplying b^c_{p} in (69) essentially serves as a correction term for an actual magnet, whose saturation magnetization slope is *not* identically zero.

The dissipation potential $D(\dot{\mathcal{H}}^r)$ in (68) is strictly convex (except for $n = +\infty$ that becomes simply convex), hence, satisfies automatically the dissipation inequality constraint. Moreover, the rate $|\dot{\mathcal{H}}^r| = \sqrt{\dot{\mathcal{H}}^r \cdot \dot{\mathcal{H}}^r}$ satisfies the material frame indifference and material symmetry conditions. By observing the experimental data on magnetic materials, we focus next on *rate-independent* ferromagnetic hysteresis responses. Thus, we consider the limiting case $n = +\infty$ at which, the dissipation potential (68) becomes $D(\dot{\mathcal{H}}^r) = b^c|\dot{\mathcal{H}}^r|$, whose derivarive with respect to $\dot{\mathcal{H}}^r$ is non-unique at $|\dot{\mathcal{H}}^r| = 0$. Hence, we start from the Legendre-Fenchel transform of D, i.e., D^* such that

$$D^*(\mathcal{B}^r) = \inf_{\dot{\mathcal{H}}^r}\left[\mathcal{B}^r \cdot \dot{\mathcal{H}}^r - b^c|\dot{\mathcal{H}}^r|\right] \tag{70}$$

in the rate-independent limit. The minimization condition of the last expression leads to a criterion known as *ferromagnetic switching surface*

$$\Phi(\mathcal{B}^r) := |\mathcal{B}^r|^2 - (b^c)^2 = 0, \tag{71}$$

which must be satisfied during the energy dissipation in a magnetic loading/unloading cycle. With (71), we rephrase the dissipation potential $D(\dot{\boldsymbol{\xi}})$ by introducing a (non-negative) Lagrange multiplier $\dot{\Lambda}$, so that

$$D(\dot{\mathcal{H}}^r) = \sup_{\mathcal{B}^r}\ \inf_{\dot{\Lambda}\geq 0}\ \left[\mathcal{B}^r \cdot \dot{\mathcal{H}}^r - \dot{\Lambda}\,\Phi(\mathcal{B}^r)\right]. \tag{72}$$

In fact, substituting $\mathcal{B}^r = b^c\dot{\mathcal{H}}^r/|\dot{\mathcal{H}}^r|$ (the minimization condition of (70)) yields exactly $D(\dot{\mathcal{H}}^r) = b^c|\dot{\mathcal{H}}^r|$ but now with a constraint (71), which must be satisfied to make the term $\dot{\Lambda}\,\Phi(\mathcal{B}^r)$ in (72) to vanish.

The constrained dissipation potential in (72) thus needs to be employed in the variational principle (28) to obtain a set of equations necessary to obtain the evolution of $\mathcal{B}^r$. These stationarity conditions of (72) are

$$\dot{\mathcal{H}}^r = \dot{\Lambda}\frac{\partial\Phi}{\partial\mathcal{B}^r}, \qquad \Phi(\mathcal{B}^r) \leq 0, \qquad \dot{\Lambda} \geq 0 \quad \text{and} \quad \dot{\Lambda}\,\Phi = 0, \tag{73}$$

where the latter three is commonly referred to be the Kraush-Kuhn-Tucker (KKT) conditions. With (73), the evolution equation for the internal variable $\mathcal{H}^r$ is now fully defined.

Remark 4.4 The limiting case of $c = 0$ leads to the energy densities associated with the non-magnetic elastomer for both, **F-H** and **F-B** models. Specifically, the condition $c = 0$ leads to the magnetic free energies (for both the models) so that

$$\rho_0\Psi^{\mathtt{H/B}}_{\mathtt{mag}} = \begin{cases} +\infty & \text{if} \quad \mathcal{H}^r \neq \mathbf{0}, \\ 0 & \text{if} \quad \mathcal{H}^r = \mathbf{0}. \end{cases} \tag{74}$$

This condition essentially constraints $\mathcal{H}^r$ to remain $\mathbf{0}$ for $c = 0$. Thus, the dissipation potential (68) vanishes and the energy densities for the **F-H** and **F-B** models read, respectively, $W^{\mathrm{H}}_{c=0} = \rho_0\Psi_{\mathtt{mech}}(I_1) - (\mu_0/2)I^{\mathrm{H}}_5$ and $W^{\mathrm{B}}_{c=0} = \rho_0\Psi_{\mathtt{mech}}(I_1) +$

$(1/2\mu_0)I_5^{\mathrm{B}}$. The limit of $c \to 1$, on the other hand, leads to the mechanically rigid hard-magnetic particle response, essentially yielding the pure magnetic switching surface model. Note, however, that in this important special case of $c \to 1$, one should replace the mechanical energy in (59) (or (60)) with one that is not infinite but instead with a large modulus in order to allow a numerical resolution of the problem.

4.8 Total Cauchy Stress in h-MREs

Although the constitutive model definitions are complete so far, the expression for the total Cauchy stress in terms of the current magnetic and mechanical variables are often sought after to gain more insight to the different stress contributions. Thus, the expressions for $\boldsymbol{\sigma}$ in terms of $\mathbb{B}$, $\mathbf{h}$, $\mathbf{b}$ and $\mathbf{h}^r$, where $\mathbb{B} = \mathbf{F}\mathbf{F}^T$ is the left Cauchy-Green tensor, in the **F**-**H** and **F**-**B** settings are provided in the following.

4.8.1 Cauchy Stress in the F-H Model

We first express W^{H} in terms of $\mathbf{F}$, $\mathbf{H}$ and $\mathcal{H}^r$ and subsequently express it to be $W^{\mathrm{H}}(\mathbf{F}, \mathbf{H}, \mathbf{h}^r) \equiv \mathrm{w}^{\mathrm{h}}(\mathbb{B}, \mathbf{h}, \mathbf{h}^r) = \rho_o \psi^{\mathrm{h}}(\mathbb{B}, \mathbf{h}, \mathbf{h}^r) - (\mu_0/2)J\mathbf{h}\cdot\mathbf{h}$, where ψ^{h} is the Helmholtz free energy density associated with the h-MRE. Moreover, we treat the Eulerian fields to be functions of $\mathbf{F}$ (or $\mathbf{R}$) and their referential (or intermediate) counterparts, such tha $\mathbf{h} = \mathbf{h}(\mathbf{F}, \mathbf{H})$ and $\mathbf{h}^r = \mathbf{h}^r(\mathbf{R}, \mathcal{H}^r)$. With these, a straightforward algebraic exercise starting from the variational statement (28) and utilizing (25), (46) leads to (see Appendix A of Mukherjee et al. (2021) for details)

$$\boldsymbol{\sigma} = \underbrace{\frac{2\rho_0}{J}\left[\frac{\partial\psi^{\mathrm{h}}}{\partial\mathbb{B}}\right]_{\mathbf{h},\mathbf{h}^r}\mathbb{B}}_{\boldsymbol{\sigma}^{\mathrm{e}}} + \underbrace{\frac{2}{J\det\mathbf{Z}}\left[\mathbf{Z}\,\mathrm{skw}\big(\mathbf{h}^r\otimes\mathbf{b}^r\big)\mathbf{V}\mathbf{Z}\right]}_{\boldsymbol{\sigma}^{\mathrm{r}}} + \underbrace{\left[\mathbf{h}\otimes\mathbf{b} - \frac{\mu_0}{2}|\mathbf{h}|^2\mathbf{I}\right]}_{\boldsymbol{\sigma}^{\mathrm{maxw}}}, \tag{75}$$

where three distinct components of the total $\boldsymbol{\sigma}$, namely the elastic $\boldsymbol{\sigma}^{\mathrm{e}}$, remanent $\boldsymbol{\sigma}^{\mathrm{r}}$ and Maxwell $\boldsymbol{\sigma}^{\mathrm{maxw}}$ stress parts are obtained. In this last expression, we introduce the Eulerian counterpart of $\mathcal{B}^r$ to be $\mathbf{b}^r = -\rho_0[\partial\psi^{\mathrm{h}}/\partial\mathbf{h}^r]_{\mathbb{B},\mathbf{h}}$, such that, $\mathbf{b}^r = \mathbf{R}\mathcal{B}^r$. Moreover, in (75) we use the explicit fourth order tensor expression for $\partial\mathbf{R}/\partial\mathbf{F}$ from (Chen & Wheeler, 1993), which, in turn, introduces the tensors $\mathbf{V}$ and $\mathbf{Z}$ defined as

$$\mathbf{V} = \mathbf{F}\mathbf{R}^T \qquad \text{and} \qquad \mathbf{Z} = \mathrm{tr}[\mathbf{V}]\mathbf{I} - \mathbf{V}. \tag{76}$$

By its very definition from $(25)_1$, where $\mathbf{S}$ is given by $(31)_2$, the total $\boldsymbol{\sigma}$ is symmetric. However, its components $\boldsymbol{\sigma}^{\mathrm{e}}$, $\boldsymbol{\sigma}^{\mathrm{r}}$ and $\boldsymbol{\sigma}^{\mathrm{maxw}}$ are not, in general, symmetric.

4.8.2 Cauchy Stress in the F-B Model

Similarly, the expression for total $\boldsymbol{\sigma}$ in the **F**-**B** model can be obtained by first expressing $W^{\mathrm{B}}(\mathbf{F}, \mathbf{B}, \boldsymbol{\mathcal{H}}^r) \equiv \mathrm{w}^{\mathrm{b}}(\mathbb{B}, \mathbf{b}, \mathbf{h}^r) = \psi^{\mathrm{b}}(\mathbb{B}, \mathbf{b}, \mathbf{h}^r) - (1/2\mu_0) J\mathbf{b} \cdot \mathbf{b}$ with the Helmholtz free energy density now expressed in terms of $\mathbb{B}$, $\mathbf{b} = \mathbf{b}(\mathbf{F}, \mathbf{B})$ and $\mathbf{h}^r = \mathbf{h}^r(\mathbf{R}, \boldsymbol{\mathcal{H}}^r)$. The expression for $\boldsymbol{\sigma}$ from the variational statement (38) and (25), (46) becomes

$$\boldsymbol{\sigma} = \underbrace{\frac{2\rho_0}{J}\left[\frac{\partial \psi^{\mathrm{b}}}{\partial \mathbb{B}}\right]_{\mathbf{b},\mathbf{h}^r} \mathbb{B}}_{\boldsymbol{\sigma}^{\mathrm{e}}} + \underbrace{\frac{2}{J \det \mathbf{Z}}\left[\mathbf{Z}\, \mathrm{skw}\left(\mathbf{h}^r \otimes \mathbf{b}^r\right) \mathbf{V}\mathbf{Z}\right]}_{\boldsymbol{\sigma}^{\mathrm{r}}} + \underbrace{\left[\mathbf{h} \otimes \mathbf{b} - \frac{\mu_0}{2}\left(|\mathbf{h}|^2 - |\mathbf{m}|^2\right)\mathbf{I}\right]}_{\boldsymbol{\sigma}^{\mathrm{maxw}}}, \tag{77}$$

where $\mathbf{b}^r = -\rho_0[\partial\psi^{\mathrm{b}}/\partial\mathbf{h}^r]_{\mathbb{B},\mathbf{b}}$. Thus, the expressions for the elastic and remanent Cauchy stresses remain the same in the **F**-**H** and **F**-**B** models, of course, the latter has a free energy density ψ^{b}, while the former has ψ^{h} in their constitutive relations. Moreover, the hydrostatic part of the Maxwell stress gets modified in the case of the **F**-**B** model, which is in agreement with the existing s-MRE constitutive models (Kankanala & Triantafyllidis, 2004; Dorfmann & Ogden, 2004; Danas, 2017).

Next, with a relative abuse in the notations,[5] we express the non-Maxwell part of $\boldsymbol{\sigma}$ to be simply the mechanical Cauchy stress contribution, so that $\boldsymbol{\sigma}^{\mathrm{mech}} = \boldsymbol{\sigma}^{\mathrm{e}} + \boldsymbol{\sigma}^{\mathrm{r}}$. Of course, not only the mechanical strains, but also the magnetic remanent fields in the h-MRE contribute to $\boldsymbol{\sigma}^{\mathrm{mech}}$. In fact, the expressions of $\boldsymbol{\sigma}^{\mathrm{r}}$ in (75) and (77) show that the remanent stress arises whenever the currnet remanent magnetization $\mathbf{h}^r$ and its dual $\mathbf{b}^r$ cease to be parallel. This particular scenario arises during the non-aligned loading of the h-MREs, leading to a "magnetic torque"-like contribution to the total $\boldsymbol{\sigma}$.

The Maxwell stress $\boldsymbol{\sigma}^{\mathrm{maxw}}$, on the other hand, remains independent of the material properties, while only depending on the local **h** and **b** fields at any point in the continuum. The mechanical and Maxwell parts of the first Piola-Kirchhoff stress can then be obtained directly via $\mathbf{S}^{\mathrm{mech}} = J\boldsymbol{\sigma}^{\mathrm{mech}}\mathbf{F}^{-T}$ and $\mathbf{S}^{\mathrm{maxw}} = J\boldsymbol{\sigma}^{\mathrm{maxw}}\mathbf{F}^{-T}$, such that $\mathbf{S} = \mathbf{S}^{\mathrm{mech}} + \mathbf{S}^{\mathrm{maxw}}$.

[5] Perhaps the best word for this term would have been the stress in the material, i.e., $\boldsymbol{\sigma}^{\mathrm{mat}}$ to distinguish it from the Maxwell part that is present even when there is no material. Nevertheless, for historical reasons we keep here the earlier notations.

5 Modeling of Isotropic Soft-MREs

In this section, we propose fully explicit, homogenization-guided phenomenological models for the s-MRE using both **F**-**H** and **F**-**B** formulations. These models have been developed independently of the previous h-MRE models and as we will show in the following are not exact duals. Instead, they are explicit and fully energetic exhibiting no dissipation. They are meant to be simpler but robust alternatives to the previous dissipative models when no (or very weak) magnetic dissipation is present.

The present phenomenological models are proposed in terms of two additional modeling parameters, which are subsequently obtained by calibration with the h-MRE models since the latter have been already calibrated with available RVE simulations. First, we propose models for incompressible MREs and then extend in an ad-hoc manner those models for quasi-incompressible ones. This extension serves only practical purposes since it allows for a simpler numerical implementation but should not be considered as an extension for arbitrarily compressible responses. In the following, we directly report the final expressions, while the reader is referred to the original work of Mukherjee et al. (2020) to find details on their derivations.

In this section, we will use only the energetic invariants, $I_{\mathtt{i}}^{\mathrm{H}}$ and $I_{\mathtt{i}}^{\mathrm{B}}$ (with $\mathtt{i} = 4, 5, 6$) defined in (52) and (53). The rest of the invariants involving the internal variable $\boldsymbol{\mathcal{H}}^r$ are not relevant in the context of s-MREs.

5.1 F-H Expressions for s-MREs

We propose a phenomenological energy function for *incompressible* s-MREs in terms of three distinct energy contributions, namely, a fully decoupled mechanical and magnetic energy and an additional coupling energy, which reads

$$W^{\mathrm{H}}(\mathbf{F}, \mathbf{H}) = \rho_0 \Psi_{\mathtt{mech}}(I_1) + \rho_0 \Psi_{\mathtt{mag}}^{\mathrm{H}}(I_5^{\mathrm{H}}) + \rho_0 \Psi_{\mathtt{couple}}^{\mathrm{H}}(I_4^{\mathrm{H}}, I_5^{\mathrm{H}}) - \frac{\mu_0}{2} I_5^{\mathrm{H}}, \qquad (78)$$

valid for all isochoric deformations states, i.e., $J = 1$. The effective mechanical energy $\Psi_{\mathtt{mech}}$ is given by (59), while the invariants I_4^{H} and I_5^{H} are those defined in (52).

Magnetic Energy

The *purely* magnetic part $\Psi_{\mathtt{mag}}^{H}$ is given in terms of a Gaussian Hypergeometric function, denoted by ${}_2\mathcal{F}_1$, as

$$\rho_0 \Psi_{\mathtt{mag}}^{\mathrm{H}}(I_5^{\mathrm{H}}) = -\frac{\mu_0\, \chi}{2} I_5^{\mathrm{H}}\, {}_2\mathcal{F}_1 \left[\frac{1}{k^{\mathrm{H}}}, \frac{2}{k^{\mathrm{H}}}, 1 + \frac{2}{k^{\mathrm{H}}}, -\left(\frac{\chi \sqrt{I_5^{\mathrm{H}}}}{m^s} \right)^{k^{\mathrm{H}}} \right]. \qquad (79)$$

In this expression, k^{H} is a positive integer that will be calibrated against numerical data. Furthermore, m^s denotes the effective saturation magnetization and $\chi \equiv \mu/\mu_0 - 1$ is the effective magnetic susceptibility (and μ the effective permeability) given by

$$m^s = c\, m^s_{\mathrm{p}}, \qquad \chi = \frac{3\, c\, \chi_{\mathrm{p}}}{3 + (1-c)\chi_{\mathrm{p}}}. \tag{80}$$

Here, χ_{p} is the magnetic susceptibility of the particles. Usually, values between 10–100 are representative of iron particles, while a value of $\chi_{\mathrm{p}} = 30$ was shown to correspond well to the commercially available carbon iron particles (CIP). This value is four times larger than that for hard magnetic NdFeB particles discussed in the previous section. The estimate for χ in (80) corresponds to the Maxwell-Garnett (or equivalently Hashin-Shtrikman) bound and was shown to be particularly accurate for isotropic s-MRE by comparison with corresponding experiments in Psarra et al. (2017) up to volume fractions of $c = 0.3$. Note further that the estimate for m^s is an exact homogenization result and has been verified by experiments in Danas et al. (2012) and numerical simulations Danas (2017) as well as shown via the approximate homoegenization estimates of Galipeau and Ponte Castañeda (2013). It implies that the effective saturation magnetization of the s-MRE composite is independent of its microstructure (e.g. whether particles are distributed isotropically or in particle-chains or have arbitrary shapes such as ellipsoids etc.) and is only a function of the volume fraction of particles c and of their individual saturation magnetization m^s_{p}. In particular, CIP have saturation magnetization much higher than that for NdFeB that attains values of $\mu_0 m^s_{\mathrm{p}} = 2.5$T.

In turn, the function ${}_2\mathcal{F}_1$ is typically expressed in terms of a series given by,

$${}_2\mathcal{F}_1[a, b, c; z] = \sum_{n=0}^{\infty} \frac{(a)_n (b)_n}{(c)_n} \frac{z^n}{n!}, \tag{81}$$

with

$$(x)_0 = 1 \qquad \text{and} \qquad (x)_n = x(x+1)\cdots(x+n-1).$$

It can be shown via rigorous convergence tests that the infinite series in (81) converge for all $z < 0$ and non-negative a, b and c (Abramowitz and Stegun 1972, pp. 81–86). Hence, (81) can be evaluated numerically in a straightforward manner (Perger et al., 1993; Hankin, 2015). Of interest, however, are the first and second derivatives of Ψ^H_{mag} with respect to $\mathbf{h}$, which, as shown in the following, take very simple algebraic forms. The derivative of the Gaussian Hypergeometric function ${}_2\mathcal{F}_1$ with respect to its argument has a very simple form, which reads

$$\mathbf{m} = -\frac{\rho_0}{\mu_0}\frac{\partial \Psi^{\mathrm{H}}_{\mathrm{mag}}}{\partial \mathbf{H}}\mathbf{F}^T = \frac{\chi \mathbf{h}}{\left[1 + (\chi)^{k^{\mathrm{H}}}\left(|\mathbf{h}|/m^s\right)^{k^{\mathrm{H}}}\right]^{1/k^{\mathrm{H}}}}. \tag{82}$$

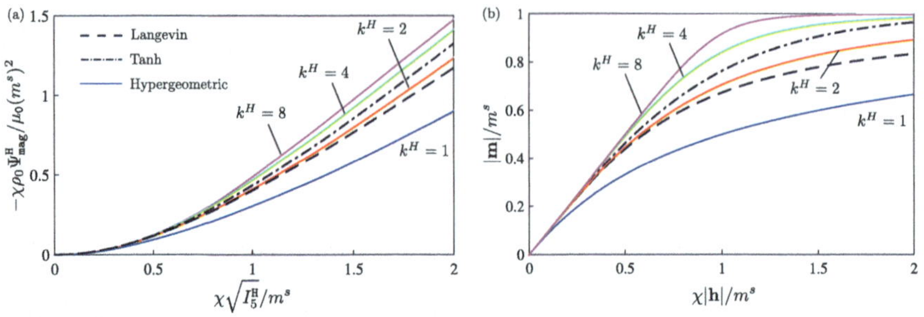

Fig. 4 Comparison of **a** the magnetic energy functions and of **b** their derivatives obtained from and hypergeometric $\Psi^{\mathrm{H}}_{\mathtt{mag}}$ saturation function given in Eq. (82) for various exponents k^{H}, the Langevin function and the hyperbolic tangent function (see (85). (Taken from Mukherjee et al. (2020))

Here, the initial susceptibility is always χ irrespective of the value of k^{H} thus leading to the same initial effective magnetization response of the s-MRE. The same is true for the saturation response, which gives $|\mathbf{m}| = m^s = c\, m^s_{\mathrm{p}}$ as required by the homogenization, experimental and numerical results. On the other hand, the rate of magnetization at moderate fields depends on the power coefficient k^{H}, which may be calibrated to follow closely available numerical or experimental data. Specifically, by direct calibration, we find in Sect. 7.1 that a value

$$k^{\mathrm{H}} = 4 \tag{83}$$

leads to a good fit for the magnetization response for all volume fractions $c \in [0, 0.3]$ and matrix shear moduli analyzed in the present study. Of course, given any alternative experimental data, a different value for k^{H} may be used. For illustration purposes, we show in Fig. 4 representative curves of the hypergeometric function and its derivative, which gives the $\mathbf{m} - \mathbf{h}$ response as evaluated from Eq. (82). For comparison, we also show magnetization curves obtained by the Langevin-type and hyberbolic tangent functions (Danas, 2017) defined by

$$\text{Langevin}: \quad \rho_0\Psi^{\mathrm{H}}_{\mathtt{mag}}(\mathbf{F}, \mathbf{H}) = -\frac{\mu_0(m^s)^2}{3\chi}\left\{\ln\left[\sinh\left(\frac{3\chi}{m^s}\sqrt{I_5^{\mathrm{H}}}\right)\right] - \ln\left[\frac{3\chi}{m^s}\sqrt{I_5^{\mathrm{H}}}\right]\right\} \tag{84}$$

$$\text{Tanh}: \quad \rho_0\Psi^{\mathrm{H}}_{\mathtt{mag}}(\mathbf{F}, \mathbf{H}) = -\frac{\mu_0(m^s)^2}{\chi}\ln\left[\cosh\left(\frac{\chi}{m^s}\sqrt{I_5^{\mathrm{H}}}\right)\right] \tag{85}$$

such that

$$\text{Langevin}: \quad \mathbf{m} = m^s\mathcal{L}\left(\frac{3\chi|\mathbf{h}|}{m^s}\right)\frac{\mathbf{h}}{|\mathbf{h}|} \tag{86}$$

$$\text{Tanh}: \quad \mathbf{m} = m^s\tanh\left(\frac{\chi|\mathbf{h}|}{m^s}\right)\frac{\mathbf{h}}{|\mathbf{h}|}. \tag{87}$$

Here, $\mathcal{L}$ denotes the Langevin function

$$\mathcal{L}(x) = \coth x - x^{-1}, \qquad x \in \mathbb{R}, \tag{88}$$

Remark 5.1 The use of a hypergeometric function is done in order to allow for flexibility in the calibration process since the homogenized response of an MRE comprising magnetic particles with Langevin-type magnetization saturation response does not lead to an effective magnetization response of a Langevin-type. Before proceeding to the coupled energy part, we note further that the decoupled mechanical and magnetic energies are expressed in terms of the homogenized material parameters, which can be evaluated directly in terms of the constituents' properties and the particle volume fraction c.

Coupled Energy

Subsequently, and similar to the h-MRE models, we express the coupled magneto-mechanical energy as a function of the two invariants I_4^{H} and I_5^{H}, thus taking the form

$$\Psi^{\mathrm{H}}_{\texttt{couple}}(I_4^{\mathrm{H}}, I_5^{\mathrm{H}}) = \Psi_4^{\mathrm{H}}(I_4^{\mathrm{H}}) - \Psi_5^{\mathrm{H}}(I_5^{\mathrm{H}}), \tag{89}$$

with

$$\rho_0 \Psi_{\texttt{i}}^{\mathrm{H}}(I_{\texttt{i}}^{\mathrm{H}}) = \beta_1^{\mathrm{H}} \frac{\mu_0 (m^s)^2}{2\chi} \ln \left[1 + \sum_{q=1}^{4} \frac{1}{c} \left(\frac{4}{5}\chi \right)^{q+1} \left(\frac{c}{\beta_2^{\mathrm{H}}} \right)^{q} \left(\frac{\sqrt{I_{\texttt{i}}^{\mathrm{H}}}}{m^s} \right)^{2q} \right], \tag{90}$$

with $\texttt{i} = 4, 5$. Following the analysis in Mukherjee et al. (2020) and modifying slightly the coefficient β_1^{H} in order to fit better with the corresponding numerical RVE simulations for $c > 0.2$, we write

$$\beta_1^{\mathrm{H}} = \frac{5g(c)\,(1-c)\chi}{16} \beta_2^{\mathrm{H}}, \tag{91}$$

with

$$g(c) = 1 + g_0 \tanh \left[\sum_{q=1}^{5} g_q c^{q+5} \right], \forall c \in [0, 0.5]$$

$$g_0, g_1, g_2, g_3, g_4, g_5 = \{2530, -3.4, 38.5, -146, 231, -132\} \tag{92}$$

and

$$\beta_2^{\mathrm{H}}(G_{\mathrm{m}}^*, c) = \alpha_1^{H}(G_{\mathrm{m}}^*) - \alpha_2^{\mathrm{H}}(G_{\mathrm{m}}^*)\mathcal{L}\left[c\, \alpha_3^{\mathrm{H}}(G_{\mathrm{m}}^*)\right], \tag{93}$$

with

$$
\begin{aligned}
&G^*_{\mathrm{m}} = G_{\mathrm{m}}/G^{\mathtt{Ref}}_{\mathrm{m}}, \quad G^{\mathtt{Ref}}_{\mathrm{m}} = 1\mathrm{MPa},\\
&\alpha^{\mathrm{H}}_1(G^*_{\mathrm{m}}) = \exp\left[-0.29\tanh\left\{0.27(\ln G^*_{\mathrm{m}}+7)\right\} - 1.575\right],\\
&\alpha^{\mathrm{H}}_2(G^*_{\mathrm{m}}) = \exp\left[4.4\mathcal{L}(-0.78\ln G^*_{\mathrm{m}}) - 5.2\right],\\
&\alpha^{\mathrm{H}}_3(G^*_{\mathrm{m}}) = \frac{0.1}{G^*_{\mathrm{m}}+0.0007} - 5.4G^*_{\mathrm{m}} + 6.75.
\end{aligned}
$$

Here, $\mathcal{L}(.)$ is the Langevin function given by (88). Specifically, the function $g(c)$ is almost unit for $c \leq 0.2$ and thus results to no changes with respect to the work of Mukherjee et al. (2020). However, in that work, it was shown that the proposed model (which was calibrated by use of the implicit homogenization model of Lefèvre et al. (2017)) tends to underestimate the coupling for $c > 0.2$ by comparison to numerical RVE simulations. The function $g(c)$ (which remains the same for the subsequent **F**-**B** formulation) serves to improve upon this issue. Also it becomes equal to zero for $c \approx 0.5$. Beyond that value the coupling is almost negligible since the mechanical response becomes substantially stiff prohibiting any magnetostriction (see corresponding results in two-dimensions Danas (2017)).

In turn, the evolution of β^{H}_2 with respect to G^*_{m} is mainly controlled by the coefficients α^{H}_1 and α^{H}_2. By contrast, the third coefficient α^{H}_3 is used to model the variation of β^{H}_2 with respect to c for a given G^*_{m}. Beyond $G^*_{\mathrm{m}} > 1$ a constant $\beta^{\mathrm{H}}_2 \approx 0.155$ is sufficient. On the other hand, for very soft, gel-like MREs, i.e., in the range of $0.001 \leq G^*_{\mathrm{m}} \leq 0.01$, the coupling coefficient β^{H}_2 becomes highly sensitive to G^*_{m} and c, resulting in a significant variation of β^{H}_2 in this particular range. A more detailed discussion on the calibration process may be found in Mukherjee et al. (2020).

Remark 5.2 The parameters β^{H}_1 and β^{H}_2 may be regarded more generally as fitting constants that may be calibrated each time to describe a specific material composition. Then, the functional form of the model is straightforward and very simple to implement since it is explicit and analytical. An important observation in this context is related to the form of the coupled energy (89), and in particular the subtraction term $\Psi^{\mathrm{H}}_4(I^{\mathrm{H}}_4) - \Psi^{\mathrm{H}}_5(I^{\mathrm{H}}_5)$. This is done for two reasons. First, the derivation of $\Psi^H_{\mathtt{couple}}$ with respect to $\mathbf{h}$ leaves the magnetization response completely unaffected at small and very large applied magnetic fields, thus allowing the hypergeometric function in Eq. (79) to completely control the $\mathbf{m}-\mathbf{h}$ response at the initial regime and the saturation regime. The second reason is that only the $I^{\mathrm{H}}_4 = \mathbf{F}^T\mathbf{h}\cdot\mathbf{F}^T\,\mathbf{h}$ part of the function contributes to the magnetostriction whenever a Eulerian field $\mathbf{h}$ is applied, while the corresponding $I^{\mathrm{H}}_5 = \mathbf{h}\cdot\mathbf{h}$ part induces no magnetostriction. Moreover, we observe that $\Psi^{\mathrm{H}}_{\mathrm{i}}$ is non-convex with respect to $\sqrt{I^{\mathrm{H}}_i}$ since its derivative increases rapidly from zero to a maximum and then gradually decreases to zero (see Fig. 5b). As we will see in the following, such a function allows to obtain a material magnetostriction response that is initially quadratic, subsequently increases in a non-quadratic manner until finally reaching a saturating state.

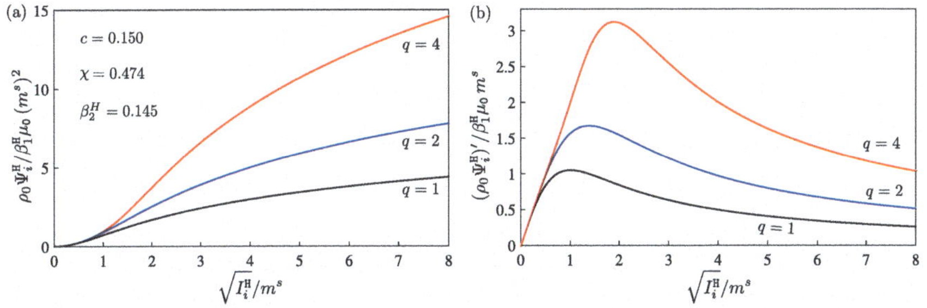

Fig. 5 Representative plots of **a** the function Ψ_i^{H} (with $\mathtt{i} = 4, 5$) and **b** of its derivative with respect to $\sqrt{I_{\mathtt{i}}^{\mathrm{H}}}$ for $q = 1, 2, 4$. (Taken from Mukherjee et al. (2020))

Quasi-incompressible Extension

In spite of the fact that the assumption of an incompressible matrix and rigid particles leads to very efficient analytical modeling of the effective response, quasi-incompressible models for the MREs are employed in most of the computational investigations due to their simplicity to incorporate them in a finite-element solver. Unfortunately, carrying out the homogenization problem for even a quasi-incompressible matrix is extremely difficult and no rigorous model is available up to date neither for the purely mechanical part nor for the magneto-mechanical part. In this regard, we propose an ad-hoc extension of the incompressible phenomenological model (78) that essentially relaxes slightly the assumption of incompressibility without affecting the aforementioned key features of the model at least in the case of high bulk modulus (i.e. quasi-incompressible materials). The proposed model reads

$$W_{\mathtt{comp}}^{\mathrm{H}}(\mathbf{F}, \mathbf{H}) = \rho_0 \Psi_{\mathtt{mech}}^{\mathtt{comp}}(\mathcal{I}_1, J) + \rho_0 \Psi_{\mathtt{mag}}^{\mathrm{H}}(I_5^{\mathrm{H}}) + \rho_0 \Psi_{\mathtt{couple}}^{\mathrm{H}}(I_4^{\mathrm{H}}, I_5^{\mathrm{H}}) - \rho_0 \frac{J\,\mu_0}{2} I_5^{\mathrm{H}}, \tag{94}$$

where $\Psi_{\mathtt{mech}}^{\mathtt{comp}}$ has already been defined in (60), while the remaining of the functions remain unchanged.

5.2 F-B Expressions for s-MREs

In principle, one may attempt to obtain an equivalent **F**-**B** model via the partial Legendre-Fenchel transformation (37) of (78) with respect to **H**. However, due to the strong nonlinearity of the functions associated with the proposed **F**-**H** model (78), one can not obtain its complementary energy in an explicit form. By contrast, in the context of h-MREs that is possible by the introduction of the internal variable, which carries all the associated nonlinearities, while the dependence of the function on the main variables **H** (or **B**) is fairly simple and thus transformable.

In any case, for completeness, we propose a complementary energy W^{B}, which has the exact same form as that of W^{H} in (78), such that

$$W^{\mathrm{B}}(\mathbf{F},\mathbf{B}) = \rho_0 \Psi_{\mathtt{mech}}(\mathbf{F}) + \rho_0 \Psi^{\mathrm{B}}_{\mathtt{mag}}(I_5^{\mathrm{B}}) + \rho_0 \Psi^{\mathrm{B}}_{\mathtt{couple}}(I_5^{\mathrm{B}}, I_6^{\mathrm{B}}) + \frac{1}{2\mu_0} I_5^{\mathrm{B}}, \qquad (95)$$

where the magneto-mechanical invariants I_5^{B} and I_6^{B} have been defined in (53). Evidently, the first term of (95) that represents the purely mechanical component of W^{B} is identical to that in (78), and is given by (59). Also, the last term of (95) represents the **F**-**B** version of the magnetostatic energy of free space (Dorfmann & Ogden, 2004).

It remains then to prescribe the two free energies, namely, the magnetic and the coupled free energy. Due to their intrinsic properties, $\Psi^{\mathrm{B}}_{\mathtt{mag}}$ and $\Psi^{\mathrm{B}}_{\mathtt{couple}}$ retain the same functional form as their **F**-**H** counterparts (79) and (89), respectively. Note that, as shown in Figs. 4 and 5, the hypergeometric ${}_2\mathcal{F}_1$ and the $\Psi^{\mathrm{H}}_{\mathtt{i}}$ functions are rich enough to model a wide variety of constitutive responses.

Magnetic Energy

In this regard, the purely magnetic part $\Psi^{\mathrm{B}}_{\mathtt{mag}}$ is chosen as

$$\Psi^{\mathrm{B}}_{\mathtt{mag}}(I_5^{\mathrm{B}}) = -\frac{\chi}{2\mu_0\,(1+\chi)} I_5^{\mathrm{B}}\, {}_2\mathcal{F}_1\left[\frac{1}{k^{\mathrm{B}}}, \frac{2}{k^{\mathrm{B}}}, 1+\frac{2}{k^{\mathrm{B}}}, -\left(\frac{\chi\sqrt{I_5^{\mathrm{B}}}}{(1+\chi)\mu_0 m^s}\right)^{k^{\mathrm{B}}}\right]. \qquad (96)$$

Similar to the **F**-**H** version, a single exponent

$$k^B = 6 \qquad (97)$$

provides a good fit to the magnetization response for all particle volume fractions and matrix shear moduli considered in this study. Note that the purely magnetic energy (96) in the **F**-**B** model is *not* an exact Legendre transform of the corresponding magnetic energy (79) of the **F**-**H** model. Thus, no direct correlation can be drawn between the model parameters k^B and k^H and their calibration values.

Coupled Energy

The coupled energy is defined by

$$\Psi^{\mathrm{B}}_{\mathtt{couple}}(I_5^{\mathrm{B}}, I_6^{\mathrm{B}}) = \Psi_6^{\mathrm{B}}(I_6^{\mathrm{B}}) - \Psi_5^{\mathrm{B}}(I_5^{\mathrm{B}}), \qquad (98)$$

with

$$\Psi_{\mathtt{i}}^{\mathrm{B}}(I_i^{\mathrm{B}}) = \beta_1^{\mathrm{B}} \frac{(1+\chi)\mu_0 (m^s)^2}{2\chi} \ln\left[1 + \sum_{q=1}^{4} \frac{1}{c}\left(\frac{5}{4}\frac{\chi}{1+\chi}\right)^{q+1}\left(\frac{c}{\beta_2^{\mathrm{B}}}\right)^q \left(\frac{\sqrt{I_{\mathtt{i}}^{\mathrm{B}}}}{\mu_0 m^s}\right)^{2q}\right], \tag{99}$$

where $\mathtt{i} = 5, 6$ and again $m^s = c\, m_{\mathtt{p}}^s$.

As in the **F**-**H** version, and following Mukherjee et al. (2020), we have

$$\beta_1^{\mathrm{B}} = \frac{16}{5} \frac{\chi(\chi+1)(1-c)}{\left([5(\chi+1) - 2(1-c)\chi^2\right]\left[(1-c)\chi^2 + 5(\chi+1)\right]} g(c) \beta_2^{\mathrm{B}}. \tag{100}$$

The function $g(c)$ is the same with that used in (92). In turn, for β_2^{B}, we use two piecewise continuous functions of $G_{\mathtt{m}}^*$ and c to model the variation of β_2^{B} in the $G_{\mathtt{m}}^* - c$ space, which reads

$$\beta_2^{\mathrm{B}}(G_{\mathtt{m}}^*, c) = \begin{cases} \alpha_1^{\mathrm{B}}(G_{\mathtt{m}}^*) - \alpha_2^{\mathrm{B}}(G_{\mathtt{m}}^*)\mathcal{L}\left[c\, \alpha_3^{\mathrm{B}}(G_{\mathtt{m}}^*)\right], & \text{if } G_{\mathtt{m}}^* \le 0.1 \\ 0.4055 - 0.5\, c\left[1 - 0.67\mathcal{L}(15 G_{\mathtt{m}}^*)\right] & \text{otherwise} \end{cases} \tag{101}$$

with

$$\alpha_1^{\mathrm{B}}(G_{\mathtt{m}}^*) = \exp\left[-0.029 \ln G_{\mathtt{m}}^* - 0.982\right],$$
$$\alpha_2^{\mathrm{B}}(G_{\mathtt{m}}^*) = \exp\left[1.78\mathcal{L}(-0.32 \ln G_{\mathtt{m}}^*) - 1.78\right],$$
$$\alpha_3^{\mathrm{B}}(G_{\mathtt{m}}^*) = \exp\left[0.14 - 0.54 \ln G_{\mathtt{m}}^*\right].$$

Here, $\mathcal{L}(.)$ is again the Langevin function defined in (88). The first function is similar to β_2^{H} with three coefficients α_1^{B}, α_2^{B} and α_3^{B}, which are functions of $G_{\mathtt{m}}^*$, whereas, the second function, which models β_2^{B} for all $G_{\mathtt{m}}^* > 0.1$, is rather a simple function of $G_{\mathtt{m}}^*$ and c. We note that the two fitting functions for β_2^B have approximately the same magnitude near $G_{\mathtt{m}}^* = 0.1$. Thus, the particular choice of piecewise continuous β_2^B ensures a constant transition from the Langevin decay to the linear decrease regime.

Remark 5.3 As stated earlier, (98) retains an identical functional form as its **F**-**H** counterpart in (89) except that the magneto-mechanical coupling is now modeled in terms of the invariant I_6^{B}. This choice of the coupling invariant is not arbitrary, rather, is directly equivalent to the **F**-**H** model. The invariant I_4^{H} can be expressed in terms of the Eulerian **h** as $I_4^{\mathrm{H}} = \mathbf{F}^T\mathbf{h} \cdot \mathbf{F}^T\mathbf{h}$. The Legendre-Fenchel transform of that invariant leads to the invariant $I_6^{\mathrm{B}} = \mathbf{F}^T\mathbf{b} \cdot \mathbf{F}^T\mathbf{b}$.

Quasi-incompressible Extension
A quasi-incompressible version of (95) is given by

$$W^{\mathrm{B}}_{\mathtt{comp}}(\mathbf{F},\mathbf{B}) = \rho_0\Psi^{\mathtt{comp}}_{\mathtt{mech}}(\mathcal{I}_1, J) + \rho_0\Psi^{\mathrm{B}}_{\mathtt{mag}}(I^{\mathrm{B}}_5) + \rho_0\Psi^{\mathrm{B}}_{\mathtt{couple}}(I^{\mathrm{B}}_5, I^{\mathrm{B}}_6) + \frac{1}{2\mu_0 J} I^{\mathrm{B}}_5, \tag{102}$$

where $\rho_0\Psi^{\mathtt{comp}}_{\mathtt{mech}}$ is given by (60).

5.3 *Total Cauchy Stress in s-MREs*

Similar to the h-MRE analysis, the expression for the total Cauchy stress in terms of the current magnetic and mechanical variables are often sought after to gain more insight to the different stress contributions. Thus, the expressions for $\boldsymbol{\sigma}$ in terms of $\mathbb{B}$, $\mathbf{h}$ and $\mathbf{b}$, where $\mathbb{B} = \mathbf{F}\mathbf{F}^T$ is the left Cauchy-Green tensor, in the **F**-**H** and **F**-**B** settings are provided in the following. The expressions obtained in the context of s-MREs may be retrieved readily from those for the h-MREs in Eqs. (75) and (77) by simply setting the remanent part equal to zero.

5.3.1 Cauchy Stress in the F-H Model

We first express W^{H} in terms of $\mathbf{F}$, $\mathbf{H}$ and subsequently express it to be $W^{\mathrm{H}}(\mathbf{F},\mathbf{H}) \equiv \mathrm{w}^{\mathrm{h}}(\mathbb{B},\mathbf{h}) = \rho_o\psi^{\mathrm{h}}(\mathbb{B},\mathbf{h}) - (\mu_0/2)J\mathbf{h}\cdot\mathbf{h}$, where ψ^{h} is the Helmholtz free energy density associated with the s-MRE. Moreover, we treat the Eulerian fields to be functions of $\mathbf{F}$ and their referential counterparts, such that $\mathbf{h} = \mathbf{h}(\mathbf{F},\mathbf{H})$. This leads to

$$\boldsymbol{\sigma} = \underbrace{\frac{2\rho_0}{J}\left[\frac{\partial\psi^{\mathrm{h}}}{\partial\mathbb{B}}\right]_{\mathbf{h}}\mathbb{B}}_{\boldsymbol{\sigma}^{\mathrm{e}}\equiv\boldsymbol{\sigma}^{\mathtt{mech}}} + \underbrace{\left[\mathbf{h}\otimes\mathbf{b} - \frac{\mu_0}{2}|\mathbf{h}|^2\mathbf{I}\right]}_{\boldsymbol{\sigma}^{\mathtt{maxw}}}, \tag{103}$$

where two distinct components of the total $\boldsymbol{\sigma}$, namely the elastic or mechanical $\boldsymbol{\sigma}^{\mathtt{mech}}$ and Maxwell $\boldsymbol{\sigma}^{\mathtt{maxw}}$ stress parts are obtained. By its very definition from $(25)_1$, where $\mathbf{S}$ is given by $(31)_2$, the total $\boldsymbol{\sigma}$ is symmetric. However, its components $\boldsymbol{\sigma}^{\mathtt{mech}}$ and $\boldsymbol{\sigma}^{\mathtt{maxw}}$ are not, in general, symmetric.

5.3.2 Cauchy Stress in the F-B Model

Similarly, the expression for total $\boldsymbol{\sigma}$ in the **F**-**B** model can be obtained by first expressing $W^{\mathrm{B}}(\mathbf{F}, \mathbf{B}) \equiv \mathrm{w}^{\mathrm{b}}(\mathbb{B}, \mathbf{b}) = \psi^{\mathrm{b}}(\mathbb{B}, \mathbf{b}) - (1/2\mu_0)J\mathbf{b} \cdot \mathbf{b}$ with the Helmholtz free energy density now expressed in terms of $\mathbb{B}$ and $\mathbf{b} = \mathbf{b}(\mathbf{F}, \mathbf{B})$. The expression for $\boldsymbol{\sigma}$ from the variational statement (38) and (25) becomes

$$\boldsymbol{\sigma} = \underbrace{\frac{2\rho_0}{J}\left[\frac{\partial \psi^{\mathrm{b}}}{\partial \mathbb{B}}\right]_{\mathbf{b}} \mathbb{B}}_{\boldsymbol{\sigma}^{\mathrm{e}} \equiv \boldsymbol{\sigma}^{\mathrm{mech}}} + \underbrace{\left[\mathbf{h} \otimes \mathbf{b} - \frac{\mu_0}{2}\left(|\mathbf{h}|^2 - |\mathbf{m}|^2\right)\mathbf{I}\right]}_{\boldsymbol{\sigma}^{\mathrm{maxw}}}. \tag{104}$$

Thus, the expressions for the elastic or mechanical and remanent Cauchy stresses remain the same in the **F**-**H** and **F**-**B** models, of course, the latter has a free energy density ψ^{b}, while the former has ψ^{h} in their constitutive relations. Moreover, the hydrostatic part of the Maxwell stress gets modified in the case of the **F**-**B** model, which is in agreement with the existing s-MRE constitutive models (Kankanala & Triantafyllidis, 2004; Dorfmann & Ogden 2004; Danas, 2017).

Again, the Maxwell stress $\boldsymbol{\sigma}^{\mathrm{maxw}}$ remains independent of the material properties, while only depending on the local **h** and **b** fields at any point in the continuum. The mechanical and Maxwell parts of the first Piola-Kirchhoff stress can then be obtained directly via $\mathbf{S}^{\mathrm{mech}} = J\boldsymbol{\sigma}^{\mathrm{mech}}\mathbf{F}^{-T}$ and $\mathbf{S}^{\mathrm{maxw}} = J\boldsymbol{\sigma}^{\mathrm{maxw}}\mathbf{F}^{-T}$, such that $\mathbf{S} = \mathbf{S}^{\mathrm{mech}} + \mathbf{S}^{\mathrm{maxw}}$.

6 Numerical Implementations for MREs

The rate-type variational principles for the **F**-**H** and **F**-**B** models in Sect. 3 are now expressed in a time discrete form to analyze periodic unit-cells or any practical boundary value problem (BVP) involving MREs. First, the scalar potential-based **F**-**H** model is presented in a time discrete form. This will be followed by the specification of the corresponding vector potential-based **F**-**B** model. We disuss the general case of h-MREs, whereby that of the s-MREs may be obtained by omitting the terms related to magnetic dissipation.

6.1 Time Discrete Variational Principle for F-H Formulation

The scalar potential-based **F**-**H** model needs to be solved for the displacement **u** and potential φ, such that $\mathbf{F} = \mathbf{I} + \mathrm{Grad}\mathbf{u}$ and $\mathbf{H} = -\mathrm{Grad}\varphi$ both satisfy the relevant Dirichlet boundary conditions. Specifically in the incremental setting of a numerical solution, we consider the state of the continuum to be known at a time t, from which we solve for the minimizing fields **u** and φ for the next time step $\tau = t + \Delta t$. We henceforth indicate all the variables with the subscripts "t" or "τ" to indicate that the variables are computed at a given discrete time.

First, the variational principle (28) upon substitution of $D(\dot{\mathcal{H}}^r)$ from (72) reads

$$\dot{\Pi}^{\mathrm{H}} = \inf_{\dot{\mathbf{u}}\in\widetilde{\mathcal{U}}} \sup_{\dot{\varphi}\in\widetilde{\mathcal{G}}} \inf_{\dot{\mathcal{H}}^r\in\mathbb{R}^3} \left[\int_{\mathbb{R}^3} \dot{W}^{\mathrm{H}}(\mathbb{C}, \mathbf{H}, \mathcal{H}^r)\, \mathrm{d}\mathcal{V}_0 - \int_{\partial\mathcal{V}_0} \mathbf{T}\cdot\dot{\mathbf{u}}\, \mathrm{d}\mathcal{S}_0 \right.$$
$$\left. + \int_{\mathcal{V}_0} \sup_{\mathcal{B}^r} \inf_{\dot{\Lambda}\geq 0} \left\{ \mathcal{B}^r\cdot\dot{\mathcal{H}}^r - \dot{\Lambda}\Phi(\mathcal{B}^r) \right\} \mathrm{d}\mathcal{V}_0 \right]. \tag{105}$$

Expressing all the rates in time discrete form like $\dot{\mathbf{u}} = (\mathbf{u}_\tau - \mathbf{u}_t)/\Delta t$ and taking note on the fact that the state at time t is already converged, one can express the rate-type variational principle (105) in a time-discrete form so that

$$\Pi^{\mathrm{H}}_\tau = \inf_{\mathbf{u}_\tau\in\mathcal{U}} \sup_{\varphi_\tau\in\mathcal{G}} \left[\int_{\mathcal{V}_0} \mathcal{W}^{\mathrm{H}}_\tau(\mathbb{C}, \mathbf{H})\, \mathrm{d}\mathcal{V}_0 + \int_{\mathbb{R}^3\setminus\mathcal{V}_0} W^{\mathrm{H}}_{c=0,\tau}(\mathbb{C}, \mathbf{H})\, \mathrm{d}\mathcal{V}_0 \right.$$
$$\left. - \int_{\partial\mathcal{V}_0} \mathbf{T}\cdot\mathbf{u}_\tau\, \mathrm{d}\mathcal{S}_0 \right]. \tag{106}$$

Here the subscript "τ" with W^{H} and $\mathcal{W}^{\mathrm{H}}$ both indicate that all their arguments are at a discrete time τ. In (106) we have introduced a *reduced energy density* $\mathcal{W}^{\mathrm{H}}_\tau$, which is, in turn, the variational principle employed for the computation for the internal variable $\mathcal{H}^r_\tau$ locally at each point of the computation domain, such that

$$\mathcal{W}^{\mathrm{H}}_\tau(\mathbb{C}, \mathbf{H}) = \inf_{\mathcal{H}^r_\tau} \sup_{\mathcal{B}^r} \inf_{\Delta\Lambda\geq 0} \left\{ W^{\mathrm{H}}_\tau(\mathbb{C}, \mathbf{H}, \mathcal{H}^r) + \mathcal{B}^r\cdot\mathcal{H}^r_\tau - \Delta\Lambda\Phi(\mathcal{B}^r) \right\}. \tag{107}$$

This last variational statement, in turn, leads to the time-discrete forms of the KKT conditions stated in (73). Finally, the admissible sets for $\mathbf{u}_\tau$ and φ_τ are given by, respectively,

$$\mathcal{U} \equiv \left\{ \mathbf{u}_\tau : \mathbf{F}_\tau = \mathbf{I} + \mathrm{Grad}\mathbf{u}_\tau,\ \forall\,\mathbf{X}\in\mathbb{R}^3,\ \mathbf{u}_\tau = \overline{\mathbf{u}}_\tau\ \forall\,\mathbf{X}\in\partial\mathcal{V}^{\mathbf{u}}_0 \right\}, \tag{108}$$

$$\mathcal{G} \equiv \left\{ \varphi_\tau : \mathbf{H}_\tau = -\mathrm{Grad}\varphi_\tau,\ \forall\,\mathbf{X}\in\mathbb{R}^3,\ \varphi_\tau = \overline{\varphi}_\tau\ \forall\,\mathbf{X}\in\partial\mathcal{V}^{\varphi}_\infty \right\}. \tag{109}$$

Thus, for an initial guess $\mathbf{u}_\tau$ and φ_τ we first update the internal variable $\mathcal{H}^r_\tau$ via extremizing (107). Then the updated $\mathcal{H}^r_\tau$ is used to compute for the corrector for $\mathbf{u}_\tau$ and φ_τ from the global implicit solver. Thus, the introduction of the reduced energy density allows us to update $\mathcal{H}^r_\tau$ locally at each integration point, while computing for $\mathbf{u}_\tau$ and φ_τ from the global variational principle. This computation algorithm provides efficient update procedure for $\mathbf{u}$, φ and $\mathcal{H}^r$ and facilitates the implementation in the commercially-available finite-element solvers like ABAQUS/Standard (Miehe et al., 2011; Rosato & Miehe, 2014; Mukherjee et al. 2021).

We note further here that corresponding expressions can be obtained for s-MREs by simply dropping the time discrete character of the previous equations, setting dissipation equal to zero and using the energy functions presented in Sect. 5.1.

6.2 Time Discrete Variational Principle for F-B Formulation

The time discrete equivalent of the **F**-**B**-based variational principle (38) can be obtained in a similar way to (106) after substituting $D(\dot{\mathcal{H}}^r)$ from (72), finally leading to

$$\Pi_\tau^{\mathrm{B}} = \inf_{\mathbf{u}_\tau \in \mathcal{U}} \inf_{\mathbf{A}_\tau \in \mathcal{B}} \left[\int_{\mathcal{V}_0} \mathcal{W}_\tau^{\mathrm{B}}(\mathbb{C}, \mathbf{B}) \, \mathrm{d}\mathcal{V}_0 + \int_{\mathbb{R}^3 \setminus \mathcal{V}_0} W_{c=0,\tau}^{\mathrm{B}}(\mathbb{C}, \mathbf{B}) \, \mathrm{d}\mathcal{V}_0 - \int_{\partial \mathcal{V}_0} \mathbf{T} \cdot \mathbf{u}_\tau \, \mathrm{d}\mathcal{S}_0 \right], \tag{110}$$

where the reduced energy density $\mathcal{W}_\tau^{\mathrm{B}}(\mathbb{C}, \mathbf{B})$ reads

$$\mathcal{W}_\tau^{\mathrm{B}}(\mathbb{C}, \mathbf{B}) = \inf_{\mathcal{H}_\tau^r} \sup_{\mathcal{B}^r} \inf_{\Delta\Lambda \geq 0} \left\{ W_\tau^{\mathrm{B}}(\mathbb{C}, \mathbf{B}, \mathcal{H}^r) + \mathcal{B}^r \cdot \mathcal{H}_\tau^r - \Delta\Lambda \Phi(\mathcal{B}^r) \right\}. \tag{111}$$

Again, the extremizaiton of (111) leads to the KKT conditions for the **F**-**B** model and thus, to the update equations for $\mathcal{H}_\tau^r$. The admissible set $\mathcal{U}$ for the displacement field remains the same as in (108), while the admissible set for the vector potential $\mathbf{A}_\tau$ reads

$$\mathcal{B} \equiv \left\{ \mathbf{A}_\tau : \mathbf{B}_\tau = \mathrm{Curl}\, \mathbf{A}_\tau, \ \mathrm{Div}\, \mathbf{A}_\tau = 0, \ \forall \mathbf{X} \in \mathbb{R}^3, \ \mathbf{A}_\tau = \overline{\mathbf{A}}_\tau, \ \forall \mathbf{X} \in \partial \mathcal{V}_0^{\mathbf{A}} \right\}, \tag{112}$$

where the condition Div $\mathbf{A}_\tau = 0$ is the well-known Coulomb gauge that leads to an uniquely defined vector potential $\mathbf{A}_\tau$ (Biro & Preis, 1989; Stark et al., 2015). The implementation of the Coulomb gauge may be done in various manners. Here, we use a penalty formulation described in Dorn et al. (2021) together with under-integration of the constraint term. Again, for an initial guess of $\mathbf{u}_\tau$ and $\mathbf{A}_\tau$, the internal variable $\mathcal{H}_t^r$ is updated to be $\mathcal{H}_\tau^r$ at the local integration points. The subsequent global increments for the $\mathbf{u}_\tau$ and $\mathbf{A}_\tau$ are carried out via using the *already updated* $\mathcal{H}_\tau^r$. The correction increments for $\mathbf{u}_\tau$ and $\mathbf{A}_\tau$ continues until a global convergence is achieved.

Again, we note that corresponding expressions can be obtained for s-MREs by simply dropping the time discrete character of the previouyes equations, setting dissipation equal to zero and using the energy functions presented in Sect. 5.2.

6.3 *The Periodic Numerical Homogenization Problem*

The above general time discrete variational principles can be easily modified to deal with a periodic problem. The main difference between a periodic and a standard boundary value problem (BVP) lies in the domain that we analyze the problem and the corresponding boundary conditions. While a BVP comprises magnetic or non-magnetic bodies and potentially an air domain that may extend far from the bodies analyzed, the periodic homogenization problem only considers *by definition* a unit-cell comprising the phases that are analyzed. By extension, there is no presence of surrounding air since the surrounding domain is filled by repetition of the principal unit-cell in all three dimensions *ad infinitum*. For that to be true, one needs to apply periodic boundary conditions while maintaining geometric periodicity of the boundary of the unit-cell (Michel et al., 1999) (although the later is not an absolutely necessary condition but mostly a convenient one). In the following, we discuss the homogenization problem in the context of h-MREs, i.e., dissipative systems using the previously presented time discrete variational formulations. Corresponding straightforward expressions can then be obtained for s-MREs by simply dropping the time discrete character of the previous formulations.

As shown in Fig. 6, each point of the macro-continuum $\mathcal{V}_0$ (Fig. 6a) is assumed to be described well at the microscale by a representative volume element (RVE) having a reference volume of $\mathcal{V}_0^{\#}$ and comprising two (or more) phases, denoted as $\mathtt{i} = \mathtt{p}, \mathtt{m}$ representing the particle and matrix phase, respectively (Fig. 6c). This assumption may be considered sufficient for the present h-MRE composites provided that the particle size is sufficiently smaller than the specimen analyzed.[6] Then, for spatially and temporally (quasi-static here) slowly varying mechanical and magnetic fields at the macroscopic scales, the previous microstructural assumptions allow for separation of length scales ($\mathcal{V}_0^{\#} \ll \mathcal{V}_0$). In addition, following Danas (2017), we consider a slowly varying microstructure, so that the microstructure can be assumed to be (locally) periodic (see Fig. 6b). This interpretation results in periodic boundary conditions applied on a single RVE (see Fig. 6c).

6.3.1 Local Energy Density of the Constituents

Henceforth, the microscopic field variables along with the energy functions and the corresponding invariants are indicated with a $\breve{(\cdot)}$ symbol in order to distinguish them from their macroscopic counterparts. Since the microstructure is heterogeneous, the referential representations of the local energy density $\breve{W}$ and the local dissipation

[6] The particle size in typical h-MREs is in the order of 10–30 μm, while a cubic RVE as we will see contains approximately five particles per direction, i.e., has a side of ∼50–150 μm at moderate volume fractions. The specimen sizes in actual experiments are usually in the centimeter scale and thus are sufficiently larger than the microstructure. In turn, such models including any type of phenomenological ones should be used with caution in the context of slender structures such as those in Psarra et al. (2017) and Kim et al. (2018), where one or two dimensions of the specimen may be a fraction of a millimeter. In that case, additional calibration may be required.

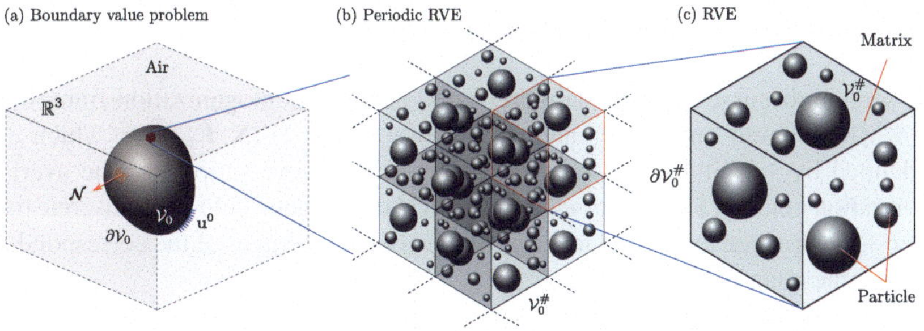

Fig. 6 Schematic diagram of **a** macroscopic boundary value problem involving a MRE sample in air having a reference volume $\mathcal{V}_0$ with unit normal $\mathcal{N}$ on the boundary $\partial\mathcal{V}_0$ and a representative boundary with fixed displacement $\mathbf{u}_0$, **b** periodic arrangement of a RVE with polydisperse spherical inclusions and **c** a RVE occupying a reference volume $\mathcal{V}_0^\#$ and boundary $\partial\mathcal{V}_0^\#$

potential $\breve{D}$ both depend on the microscopic (here) reference coordinate $\mathbf{X}$.[7] Also for brevity in the presentation, we will use the notation introduced in (44) to denote simultaneously whenever possible with a unique symbol $\breve{\mathbf{G}} = \{\breve{\mathbf{H}}, \breve{\mathbf{B}}\}$. Thus, one has

$$\breve{W}(\mathbf{X}, \breve{\mathbf{F}}, \breve{\mathbf{G}}, \breve{\mathcal{H}}^r) = \Theta(\mathbf{X})\breve{W}_{\mathrm{m}}^{\mathrm{G}}(\breve{\mathbf{F}}, \breve{\mathbf{G}}, \breve{\mathcal{H}}^r) + (1 - \Theta(\mathbf{X}))\breve{W}_{\mathrm{p}}^{\mathrm{G}}(\breve{\mathbf{F}}, \breve{\mathbf{G}}, \breve{\mathcal{H}}^r) \quad (113)$$

and

$$\breve{D}(\mathbf{X}, \dot{\breve{\mathcal{H}}}^r) = \Theta(\mathbf{X})\breve{D}_{\mathrm{m}}(\dot{\breve{\mathcal{H}}}^r) + (1 - \Theta(\mathbf{X}))\breve{D}_{\mathrm{p}}(\dot{\breve{\mathcal{H}}}^r), \quad (114)$$

In this last two expressions, $\Theta(\mathbf{X})$ denotes the characteristic function taking the value $\Theta(\mathbf{X}) = 1$ if $\mathbf{X} \in \mathcal{V}_0^{\#\mathrm{m}}$ and $\Theta(\mathbf{X}) = 0$ if $\mathbf{X} \in \mathcal{V}_0^{\#\mathrm{p}}$. The microscopic energy densities $\breve{W}_{\mathrm{p}}^{\mathrm{G}}$ and $\breve{W}_{\mathrm{m}}^{\mathrm{G}}$ are directly identified with the corresponding ones discussed in the previous sections by simply taking the limits $c = 1$ for the particle phase and $c = 0$ for the matrix phase, while all field variables $\breve{\mathbf{F}}$, $\breve{\mathbf{G}}$ and $\breve{\mathcal{H}}^r$ should be simply replaced by the corresponding overscript ones $\breve{(\cdot)}$.

Remark 6.1 In (114), we may readily set $\breve{D}_{\mathrm{m}} = 0$ since the magnetic dissipation is identically zero in the non-magnetic polymer matrix in the present analysis. Nonetheless, note that other microstructures of more phases or different properties may be readily analyzed by the present approach. For instance, one may have a magnetic polymer together with magnetic particles of different properties or anything else that one may consider useful to analyze. In this sense, the above descriptions serve only as a representative example.

---

[7] In principle, one may introduce a different notation for the position vector to insist that it corresponds to the reference coordinate in the microscopic scale which is obviously a different measure from the reference coordinate in the macroscopic scale, i.e., $\breve{\mathbf{X}} \equiv \mathbf{X}$. Nonetheless for the sake of simplicity in notation, we drop the superscript from $\mathbf{X}$, since all calculation regarding a unit-cell will always take place in the microscopic scale.

6.3.2 Incremental Homogenization Framework

This section discusses briefly the incremental periodic homogenization framework for h-MREs based on an incremental micro-potential $\breve{\mathcal{W}}(\mathbf{X}, \breve{\mathbf{F}}_\tau, \breve{\mathbf{G}}_\tau)$, which can be defined by substituting all field quantities with an overscript $\breve{(\cdot)}$. The average deformation gradient $\mathbf{F}$ and Lagrangian h-field $\mathbf{H}$ or magnetic field $\mathbf{B}$ at a discrete time $\tau \equiv t + \Delta t$ are then expressed in terms of the volume averages of the corresponding microscopic quantities, so that

$$\mathbf{F}_\tau = \frac{1}{|\mathcal{V}_0^\#|} \int_{\mathcal{V}_0^\#} \breve{\mathbf{F}}_\tau(\mathbf{X}) \, \mathrm{d}V, \qquad \mathbf{G}_\tau = \frac{1}{|\mathcal{V}_0^\#|} \int_{\mathcal{V}_0^\#} \breve{\mathbf{G}}_\tau(\mathbf{X}) \, \mathrm{d}V, \tag{115}$$

respectively.

The microscopic displacements $\breve{\mathbf{u}}_\tau(\mathbf{X})$, the microscopic scalar potential $\breve{\varphi}_\tau(\mathbf{X})$ and vector potential $\breve{\mathbf{A}}_\tau(\mathbf{X})$ are additively decomposed into linear (macroscopic) and higher order (microscopic fluctuation) contributions

$$\begin{cases} \breve{\mathbf{u}}_\tau(\mathbf{X}) = (\mathbf{F}_\tau - \mathbf{I}) \cdot \mathbf{X} + \widetilde{\mathbf{u}}_\tau(\mathbf{X}), \\ \breve{\varphi}_\tau(\mathbf{X}) = -\mathbf{H}_\tau \cdot \mathbf{X} + \widetilde{\varphi}_\tau(\mathbf{X}) \qquad\qquad \forall\, \mathbf{X} \in \mathcal{V}_0^\# \\ \breve{\mathbf{A}}_\tau(\mathbf{X}) = \dfrac{1}{2}\mathbf{B} \times \mathbf{X} + \widetilde{\mathbf{A}}_\tau(\mathbf{X}), \end{cases} \tag{116}$$

where $\widetilde{\mathbf{u}}_\tau(\mathbf{X})$, $\widetilde{\varphi}_\tau(\mathbf{X})$ and $\widetilde{\mathbf{A}}_\tau(\mathbf{X})$ are the relevant periodic (with periodicity that of the unit-cell) fluctuation fields. Their average over $\mathcal{V}_0^\#$ is required to vanish such that (116) is consistent with (115), which is automatically fulfilled for $\mathcal{V}_0^\#$*-periodic* fluctuation fields.

In the dissipative problem, one then has to solve first the local minimization problem with respect to the internal variable such that we can define the local energy (following exactly the definition (107) or (111))

$$\breve{\mathcal{W}}_\tau^{\mathrm{G}}(\mathbf{X}, \breve{\mathbf{F}}_\tau, \breve{\mathbf{G}}_\tau) = \inf_{\breve{\mathcal{H}}^r} \sup_{\breve{\mathcal{B}}^r} \inf_{\Delta\breve{\Lambda} \geq 0} \Big\{ \breve{W}_\tau^{\mathrm{G}}(\mathbf{X}, \breve{\mathbf{F}}, \breve{\mathbf{H}}, \breve{\mathcal{H}}^r) + \\ [1ex] \breve{\mathcal{B}}^r \cdot \breve{\mathcal{H}}^r{}_\tau - \Delta\breve{\Lambda}\, \Phi(\mathbf{X}, \breve{\mathcal{B}}^r) \Big\}. \tag{117}$$

As a consequence, the incremental homogenized energy $\mathcal{W}_\tau^{\mathrm{G}}$ reads

$$\mathcal{W}_\tau^{\mathrm{G}}(\mathbf{F}_\tau, \mathbf{G}_\tau) = \inf_{\breve{\mathbf{u}}_\tau \in \mathcal{K}(\mathbf{F}_\tau)} \sup_{\{\breve{\varphi}_\tau | \breve{\mathbf{A}}_\tau\} \in \mathcal{G}(\mathbf{G}_\tau)} \left[\frac{1}{\mathcal{V}_0^\#} \int_{\mathcal{V}_0^\#} \breve{\mathcal{W}}_\tau^{\mathrm{G}}(\mathbf{X}, \breve{\mathbf{F}}_\tau, \breve{\mathbf{G}}_\tau) \, \mathrm{d}V \right], \tag{118}$$

where $\mathcal{K}$ and $\mathcal{G}$ represents the sets of admissible microscopic displacement and magnetic scalar or vector potential fields, defined, respectively, as

$$\mathcal{K}(\mathbf{F}_{\mathcal{T}}) = \left\{\breve{\mathbf{F}}_{\mathcal{T}} = \mathbf{I} + \text{Grad}\,\breve{\mathbf{u}}_{\mathcal{T}}, \breve{\mathbf{u}}_{\mathcal{T}} = (\mathbf{F}_{\mathcal{T}} - \mathbf{I}) \cdot \mathbf{X} + \widetilde{\mathbf{u}}_{\mathcal{T}},\ \widetilde{\mathbf{u}}_{\mathcal{T}} \text{ periodic in } \mathcal{V}_0^{\#}\right\} \quad (119)$$

and

$$\begin{cases} \text{F} - \text{H case}: \\ \mathcal{G}(\mathbf{H}_{\mathcal{T}}) = \left\{\breve{\mathbf{H}}_{\mathcal{T}} = -\text{Grad}\,\breve{\varphi}_{\mathcal{T}}, \breve{\varphi}_{\mathcal{T}} = -\mathbf{H}_{\mathcal{T}} \cdot \mathbf{X} + \widetilde{\varphi}_{\mathcal{T}},\ \widetilde{\varphi}_{\mathcal{T}} \text{ periodic in } \mathcal{V}_0^{\#}\right\} \\ \text{F} - \text{B case}: \\ \mathcal{G}(\mathbf{B}_{\mathcal{T}}) = \left\{\breve{\mathbf{B}}_{\mathcal{T}} = \text{Curl}\,\breve{\mathbf{A}}_{\mathcal{T}}, \breve{\mathbf{A}}_{\mathcal{T}} = \dfrac{1}{2}\mathbf{B}_{\mathcal{T}} \times \mathbf{X} + \widetilde{\mathbf{A}}_{\mathcal{T}},\ \widetilde{\mathbf{A}}_{\mathcal{T}} \text{ periodic in } \mathcal{V}_0^{\#}\right\} \end{cases} \quad (120)$$

Applying then the Hill-Mandel lemma, we obtain the homogenized constitutive relations defined as

$$\mathbf{S}_{\mathcal{T}} = \frac{\partial \mathcal{W}_{\mathcal{T}}^{\text{G}}}{\partial \mathbf{F}_{\mathcal{T}}}(\mathbf{F}_{\mathcal{T}}, \mathbf{G}_{\mathcal{T}}), \quad \begin{cases} \text{F} - \text{H case}: \ \mathbf{B}_{\mathcal{T}} = -\dfrac{\partial \mathcal{W}_{\mathcal{T}}^{\text{H}}}{\partial \mathbf{H}_{\mathcal{T}}}(\mathbf{F}_{\mathcal{T}}, \mathbf{H}_{\mathcal{T}}), \\ \text{F} - \text{B case}: \ \mathbf{H}_{\mathcal{T}} = \dfrac{\partial \mathcal{W}_{\mathcal{T}}^{\text{B}}}{\partial \mathbf{B}_{\mathcal{T}}}(\mathbf{F}_{\mathcal{T}}, \mathbf{B}_{\mathcal{T}}). \end{cases} \quad (121)$$

At this stage, the definition of the incremental homogenization problem for h-MREs is formally complete and one could proceed to compare the explicit model with the numerical RVE homogenized response (118). Nevertheless, it has been shown in Danas (2017) that such use of (118) does not reveal properly the effective magneto-mechanical response that arises from interactions between the magnetic particles. In fact, it was shown that even for a non-magnetic material, one would obtain magnetostrictive strains if a magnetic field was applied. This discussion is rather less straightforward and the reader is referred to the original article for more details. In view of this observation, further modifications to the incremental variational principle are necessary, as detailed in the following.

6.3.3 Augmented F-H Potential Energy for RVE Simulations

Recent works of Keip and Rambausek (2016), Danas (2017) and Mukherjee et al. (2020) pointed out a key difference between the electro-active and magneto-active boundary value problems. Electro-active elastomers are typically loaded by electrodes that are directly attached to the material. In contrast, the MREs are usually immersed in the magnetic field created by fixed poles of electromagnets that rest at a certain distance away from the MRE sample (Bodelot et al., 2017; Zhao et al., 2019). One of the main differences between those two problems is that in the first the electric fields are zero outside the body, implying a zero Maxwell stress in vacuum, while in the second the magnetic fields and thus the Maxwell stress are not zero. In an effort to appropriately take into account the pure magneto-mechanical coupling in the RVE, free from the effect of those macroscopic boundary conditions, Danas (2017) and Mukherjee et al. (2020) proposed an augmented potential energy that involves three additional loading terms to deal with the surrounding RVE medium, the applied Eulerian magnetic field and the potential control of an average mechanical stress field. This potential energy allows to describe properly the magnetic effects (including the Maxwell stresses) exerted by the surrounding RVEs on the RVE under study and is briefly revisited here for completeness. The reader is referred to (Danas, 2017) for a complete discussion on this highly non-trivial matter.

Specifically, the first additional term serves to describe the application of the *current* macroscopic h-field, $\mathbf{h}^{\mathrm{app}}$, at the level of the RVE, instead of the referential one, $\mathbf{H}$. This may be achieved by the use of a penalty term $\frac{\mu_0}{2\zeta}|\mathbf{F}_\mathcal{T}^{-T}\mathbf{H}_\mathcal{T}-\mathbf{h}_\mathcal{T}^{\mathrm{app}}|^2$ with $\zeta \ll 1$. Next, the macroscopic background energy $-\mu_0 I_5^{\mathrm{H}}/2$ (or $-\mu_0 J I_5^{\mathrm{H}}/2$ in the quasi-incompressible case) is subtracted from (118). This accounts for the presence of the neighboring RVEs (see Fig. 6b) by imposing the continuity of the macroscopic Maxwell stresses between neighboring RVEs, far from the boundaries of the specimen. Finally, to be able to prescribe macroscopic mechanical stress $\mathbf{S}_\mathcal{T}^{\mathrm{mech}}$ instead of deformation $\mathbf{F}$, one may consider the term $\mathbf{S}_\mathcal{T}^{\mathrm{mech}} : (\mathbf{F}_\mathcal{T}-\mathbf{I})$. Assembling these three additional terms together, we obtain the augmented potential energy (Mukherjee et al., 2020)

$$\begin{aligned}\mathscr{P}_\mathcal{T}^{\mathrm{H}}(\mathbf{F}_\mathcal{T},\mathbf{H}_\mathcal{T}) &= \mathcal{W}_\mathcal{T}^{\mathrm{H}}(\mathbf{F}_\mathcal{T},\mathbf{H}_\mathcal{T}) + \frac{\mu_0}{2}\mathbf{F}_\mathcal{T}^{-T}\mathbf{H}_\mathcal{T}\cdot\mathbf{F}_\mathcal{T}^{-T}\mathbf{H}_\mathcal{T} \\ &+ \frac{\mu_0}{2\zeta}|\mathbf{F}_\mathcal{T}^{-T}\mathbf{H}_\mathcal{T}-\mathbf{h}_\mathcal{T}^{\mathrm{app}}|^2 - \mathbf{S}_\mathcal{T}^{\mathrm{mech}}\cdot(\mathbf{F}_\mathcal{T}-\mathbf{I}),\end{aligned} \tag{122}$$

with $\mathcal{W}_\mathcal{T}^{\mathrm{H}}$ defined by (118) by replacing $\mathbf{G}\equiv\mathbf{H}$. The resulting Euler-Lagrange equations of the RVE response under the prescribed magnetic and mechanical loads introduced in (122) are obtained by setting $\delta\mathscr{P}_\mathcal{T}^{\mathrm{H}}(\mathbf{F}_\mathcal{T},\mathbf{H}_\mathcal{T})=0$, which leads to

$$\mathbf{S}_\mathcal{T}-\mathbf{S}_\mathcal{T}^{\mathrm{maxw}}-\mathbf{S}_\mathcal{T}^{\mathrm{mech}}=\mathbf{0},\;\; \mathbf{B}_\mathcal{T}-\mu_0\mathbb{C}^{-1}\mathbf{H}_\mathcal{T}-\frac{\mu_0}{\zeta}\mathbf{F}_\mathcal{T}^{-1}(\mathbf{F}_\mathcal{T}^{-T}\mathbf{H}_\mathcal{T}-\mathbf{h}_\mathcal{T}^{\mathrm{app}})=\mathbf{0}. \tag{123}$$

Here, $\mathbf{S}^{\mathrm{maxw}}=J\boldsymbol{\sigma}^{\mathrm{maxw}}\mathbf{F}^{-T}$ is the 1st Piola-Kirchhoff expression for the energetic Maxwell stress given in terms of $\mathbf{h}$ and $\mathbf{b}$ as defined by (75). In turn, by writing the second equation in (123) in terms of the Eulerian parts as

$$\mathbf{b}_\mathcal{T}-\mu_0\mathbf{h}_\mathcal{T}-\frac{\mu_0}{\zeta}(\mathbf{h}_\mathcal{T}-\mathbf{h}_\mathcal{T}^{\mathrm{app}})=0, \tag{124}$$

one simply obtains the magnetization constitutive relation (5), with

$$\frac{1}{\zeta}(\mathbf{h}_\mathcal{T}-\mathbf{h}_\mathcal{T}^{\mathrm{app}})=\mathbf{m}. \tag{125}$$

This is achieved since $\mathbf{h}_{\mathcal{T}} \rightarrow \mathbf{h}_{\mathcal{T}}^{\mathrm{app}}$ as $\zeta \rightarrow 0$ making the first term finite and equal to $\mathbf{m}$. Again, the reader is referred to Danas (2017) for more details on this part.

Remark 6.2 These additional terms to the homogenization variational principle *do not* alter the microscopic constitutive models and thus, the effective incremental energy $\mathcal{W}_{\mathcal{T}}^{\mathrm{H}}$, but changes the boundary conditions applied across the periodic RVE in terms of the macroscopic fields $\mathbf{S}_{\mathcal{T}}$ and $\mathbf{H}_{\mathcal{T}}$. Such a modification allows us to obtain the magneto-mechanical coupling effects arising from the local particle interactions subject to background average Eulerian magnetic fields.

6.3.4 Augmented F-B Potential Energy for RVE Simulations

In an exactly similar fashion, one may obtain an augmented potential energy in the **F**-**B** space. Specifically, the first additional term serves to describe the application of the *current* macroscopic b-field, $\mathbf{b}^{\mathrm{app}}$, at the level of the RVE, instead of the referential one, $\mathbf{B}$. This may be achieved by the use of a penalty term $\dfrac{J}{2\zeta\,\mu_0}|J^{-1}\mathbf{F}_{\mathcal{T}}\mathbf{H}_{\mathcal{T}} - \mathbf{b}_{\mathcal{T}}^{\mathrm{app}}|^2$ with $\zeta \ll 1$. Next, the macroscopic background energy $I_5^{\mathrm{B}}/2\mu_0$ (or $J\,I_5^{\mathrm{B}}/2\mu_0$ in the quasi-incompressible case) is subtracted from (118). This accounts for the presence of the neighboring RVEs (see Fig. 6b) by imposing the continuity of the macroscopic Maxwell stresses between neighboring RVEs, far from the boundaries of the specimen. Finally, to be able to prescribe macroscopic mechanical stress $\mathbf{S}_{\mathcal{T}}^{\mathrm{mech}}$ instead of deformation $\mathbf{F}$, one may consider the term $\mathbf{S}_{\mathcal{T}}^{\mathrm{mech}} : (\mathbf{F}_{\mathcal{T}} - \mathbf{I})$. Assembling these three additional terms together, we obtain the augmented potential energy (Danas, 2017)

$$\mathscr{P}_{\mathcal{T}}^{\mathrm{B}}(\mathbf{F}_{\mathcal{T}}, \mathbf{B}_{\mathcal{T}}) = \mathcal{W}_{\mathcal{T}}^{\mathrm{B}}(\mathbf{F}_{\mathcal{T}}, \mathbf{B}_{\mathcal{T}}) - \frac{1}{2\mu_0 J}\mathbf{F}_{\mathcal{T}}\mathbf{B}_{\mathcal{T}} \cdot \mathbf{F}_{\mathcal{T}}\mathbf{B}_{\mathcal{T}} + \frac{J}{2\zeta\mu_0}|J^{-1}\mathbf{F}_{\mathcal{T}}\mathbf{B}_{\mathcal{T}} - \mathbf{b}_{\mathcal{T}}^{\mathrm{app}}|^2 - \mathbf{S}_{\mathcal{T}}^{\mathrm{mech}} \cdot (\mathbf{F}_{\mathcal{T}} - \mathbf{I}), \tag{126}$$

with $\mathcal{W}_{\mathcal{T}}^{\mathrm{B}}$ defined by (118) by replacing $\mathbf{G} \equiv \mathbf{B}$. The resulting Euler-Lagrange equations of the RVE response under the prescribed magnetic and mechanical loads introduced in (126) are obtained by setting $\delta\mathscr{P}_{\mathcal{T}}^{\mathrm{B}}(\mathbf{F}_{\mathcal{T}}, \mathbf{B}_{\mathcal{T}}) = 0$, which leads to

$$\mathbf{S}_{\mathcal{T}} - \mathbf{S}_{\mathcal{T}}^{\mathrm{maxw}} - \mathbf{S}_{\mathcal{T}}^{\mathrm{mech}} = \mathbf{0}, \; \mathbf{H}_{\mathcal{T}} - \frac{1}{\mu_0 J}\mathbb{C}\mathbf{B}_{\mathcal{T}} + \frac{1}{\zeta\mu_0}\mathbf{F}_{\mathcal{T}}^{T}(J^{-1}\mathbf{F}_{\mathcal{T}}\mathbf{B}_{\mathcal{T}} - \mathbf{b}_{\mathcal{T}}^{\mathrm{app}}) = \mathbf{0}. \tag{127}$$

Here, $\mathbf{S}^{\mathrm{maxw}} = J\boldsymbol{\sigma}^{\mathrm{maxw}}\mathbf{F}^{-T}$ is the 1st Piola-Kirchhoff expression for the energetic Maxwell stress given in terms of $\mathbf{h}$ and $\mathbf{b}$ as defined by (77). In turn, by writing the second equation in (127) in terms of the Eulerian parts as

$$\mathbf{h}_{\mathcal{T}} - \frac{1}{\mu_0}\mathbf{b}_{\mathcal{T}} + \frac{1}{\zeta\mu_0}(\mathbf{b}_{\mathcal{T}} - \mathbf{b}_{\mathcal{T}}^{\mathrm{app}}) = 0, \tag{128}$$

one simply obtains the magnetization constitutive relation (5), with

$$\frac{1}{\zeta\mu_0}(\mathbf{b}_\tau - \mathbf{b}_\tau^{\mathtt{app}}) = \mathbf{m}. \tag{129}$$

Again, the reader is referred to Danas (2017) for more details on this part.

7 Results: Periodic RVE Simulations and Model Assessment

This section discusses the model assessment via comparisons with the corresponding numerical RVE results under coupled magneto-mechanical loading conditions. In all subsequent results, we use a standard incompressible Neo-Hookean energy for the polymer matrix phase, i.e.,

$$\rho_0\Psi_{\mathtt{m}}^{\mathtt{mech}}(I_1) = \frac{G_{\mathtt{m}}}{2}(I_1 - 3). \tag{130}$$

This functional form is used in (59) to obtain the effective mechanical energy for the analytical model. We recall that in the analytical model the mechanical response of the particles is considered rigid. In turn, the magnetic properties of the particle are reported in Table 2 and correspond to a commercially available NdFeB material. In particular, these parameters are obtained by fitting the purely magnetic model with the experimentally measured hysteresis loops of magnetically isotropic NdFeB particles reported in Deng et al. (2015). Evidently, the model is general enough to be able to deal with any other type of hard magnetic particles.

The numerical simulations use the same functions and parameters as the analytical model with only two differences that do not affect, however, the validity of the comparison. The first difference is the use of a quasi-incompressible energy for both the matrix and the particle phase, which is simply obtained by adding compressible terms in (59), such that it becomes

$$\rho_0\Psi_{\mathtt{comp,i}}^{\mathtt{mech}}(I_1, J) = \frac{G_{\mathtt{i}}}{2}(I_1 - 3 - 2\ln J) + \frac{G'_{\mathtt{i}}}{2}(J-1)^2, \qquad \mathtt{i} = \mathtt{m}, \mathtt{p}. \tag{131}$$

The quasi-incompressible character of the matrix is ensured by setting $G'_{\mathtt{m}} = 500G_{\mathtt{m}}$. Use of higher values has shown practically no difference in the simulated effective

Table 2 Magnetic properties of the NdFeB particles

$\chi_{\mathtt{p}}^{e}$	$\chi_{\mathtt{p}}^{r}$	$\mu_0 m_{\mathtt{p}}^{s}$ (T)	$b_{\mathtt{p}}^{c}$ (T)	μ_0 (μN · A^2)
0.105	8.0	0.842	1.062	$4\pi 10^{-1}$

results. The second difference is the use of finite but large Lamé moduli for the particle, i.e., $G_{\mathrm{p}} = 500G_{\mathrm{m}}$ and $G'_{\mathrm{p}} = 500G_{\mathrm{p}}$. The phase contrast ratio of $G_{\mathrm{p}}/G_{\mathrm{m}} = 500$ has been shown in several earlier studies (see for instance Idiart et al. (2006a), Lopez-Pamies et al. (2013), Papadioti et al. (2016)) to be sufficiently large and thus render a nearly rigid mechanical response for the particle.

The RVE geometries are constructed by use of the RSA (random sequential adsorption) method. This allows to add sequentially inclusions of spherical (Lopez-Pamies et al., 2013) or ellipsoidal (Anoukou et al., 2018) shape in a cubic periodic unit cell. We perform the calculation by ranging the number of particles from 60 to 300. Three or more realizations are used to obtain an average and scatter response of those RVEs.

The results in the following sections consider the variation of two parameters, namely, the particle volume fraction c and the matrix shear modulus G_{m}, which are, in fact, the two critical parameters that can be varied during the fabrication of h-MREs. Instead, the magnetic properties of the particle phase are kept constant.

7.1 h-MRE Models Versus FE Simulations

7.1.1 Cyclic Magnetic Loading and Calibration of β Parameter

In this section, the goal is to calibrate the coupling parameter β introduced in the coupled magneto-mechanical energy (89) and (98) for the analytical models by use of corresponding RVE simulations. For this purpose, we fix the matrix shear modulus to $G_{\mathrm{m}} = 0.5$ MPa and vary the particle volume fraction $c = 0.1$, 0.2 and 0.3. The proposed shear modulus resembles closely that of the moderately-soft, commercially-available Sylgard-184 PDMS elastomer (Park et al., 2018; Wang et al., 2019).

We consider symmetric cyclic magnetic loading in terms of $\mathbf{h}^{\mathrm{app}} = h_1^{\mathrm{app}}\mathbf{e}_1$ with a maximum amplitude $h_1^{\mathrm{app}} = 3\, m_{\mathrm{p}}^s$. Note that the loading rate does *not* play any role here, since both the macro and microscopic h-MRE models are rate-independent. As mechanical boundary conditions we employ

$$S_{11}^{\mathrm{mech}} = S_{22}^{\mathrm{mech}} = S_{33}^{\mathrm{mech}} = 0, \qquad F_{ij} = 0, \;\; \forall i \neq j. \tag{132}$$

Similar to the numerical RVE results of non-hysteretic s-MREs (Mukherjee et al., 2020), in h-MREs too, the effective magnetostriction response exhibits a certain variance with respect to the RVE realizations, even for sufficiently large number of polydisperse spherical inclusions. For the effective RVE half-cycle responses shown in Fig. 7, we employ five different RVE realizations per particle volume fraction $c = 0.1$, 0.2 and 0.3. The corresponding average magnetization, m_1/m_{p}^s and the parallel, $\lambda_1 - 1$ and transverse, $\lambda_{2,3} - 1$, magnetostrictions are shown in Fig. 7. The light-colored patches around the respective averages indicate the scatter resulting from the considered realizations.

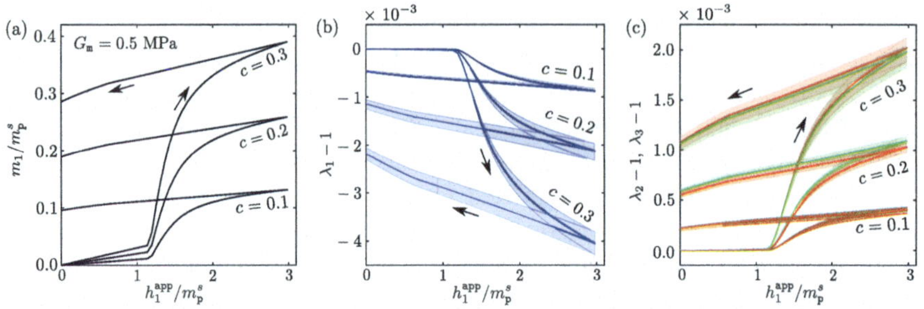

Fig. 7 Numerically computed effective **a** magnetization, **b** parallel and **c** transverse magnetostrictions the h-MRE RVEs, subjected to uniaxial Eulerian $\mathbf{h}^{\text{app}} = h_1^{\text{app}}\mathbf{e}_1$ loading/unloading. The average effective responses (solid lines) along with the range of their fluctuations (light patches) for different realizations of the respective RVEs are indicated. The RVEs of different volume fractions are comprised of random polydisperse spherical hard-magnetic inclusions, coming from three distinct families. (Taken from Mukherjee et al. (2021))

Specifically, in Fig. 7a, the scatter of the magnetization response is found to be vanishingly small, whereas, those of the parallel (λ_1 in Fig. 7b) and transverse magnetostrictions (λ_2 and λ_3 in Fig. 7c) are gradually increasing with the magnetic load. Notice from Fig. 7 that neither the effective magnetization, nor the magnetostriction saturates at higher h-fields. Rather, they maintain a slope with the applied h_1^{app}. Such response can be attributed to the inherent non-saturating magnetization response of the NdFeB particles, as observed in Fig. 3 for $\chi_{\text{p}}^e > 0$. Moreover, we observe that the overall amplitude of the magnetostriction is rather small ($\sim 10^{-3}$) indicating that a matrix with shear modulus $G_{\text{m}} = 0.5$ MPa is rather stiff in relation to the magnetic particle-to-particle forces. Even so, a permanent deformation is obtained upon complete removal of the applied magnetic field. This is obviously a direct consequence of the permanent magnetization of the particles and of their mutual interaction once magnetized permanently.

In Fig. 8a–c, we show the contours of the microscopic $\breve{b}_1/\mu_0 m_{\text{p}}^s$ fields after the end of the initial half-cycle (i.e., final state shown in Fig. 7) in the deformed RVEs for the three particle volume fractions under consideration. Figure 8a–c also shows that the magnetic self-fields under no applied $\mathbf{h}^{\text{app}}$ become considerably stronger with increasing volume fraction. Furthermore, the contours of microscopic $\breve{m}_1/m_{\text{p}}^s$ are shown in Fig. 8d–f, where we observe $\breve{m}_1 \approx m_{\text{p}}^s$ in the particles, while $\breve{m}_1 = 0$ in the non-magnetic matrix phase. In accord with the computed effective magnetostrictions in Fig. 7b, c, we observe very small overall deformation of the RVEs, although the local (microscopic) strain fields may be much higher (twice as large) and varying extensively in the matrix phase. For instance, the contours of the local nominal strain $\breve{\lambda}_1 - 1$ fields are shown in Fig. 8g–i.

The magnetization and magnetostriction response under a fully reversed proportional loading is then computed for $c = 0.1$, 0.2 and 0.3 by considering *a single* RVE of each volume fraction. These RVEs are selected to be those, whose effec-

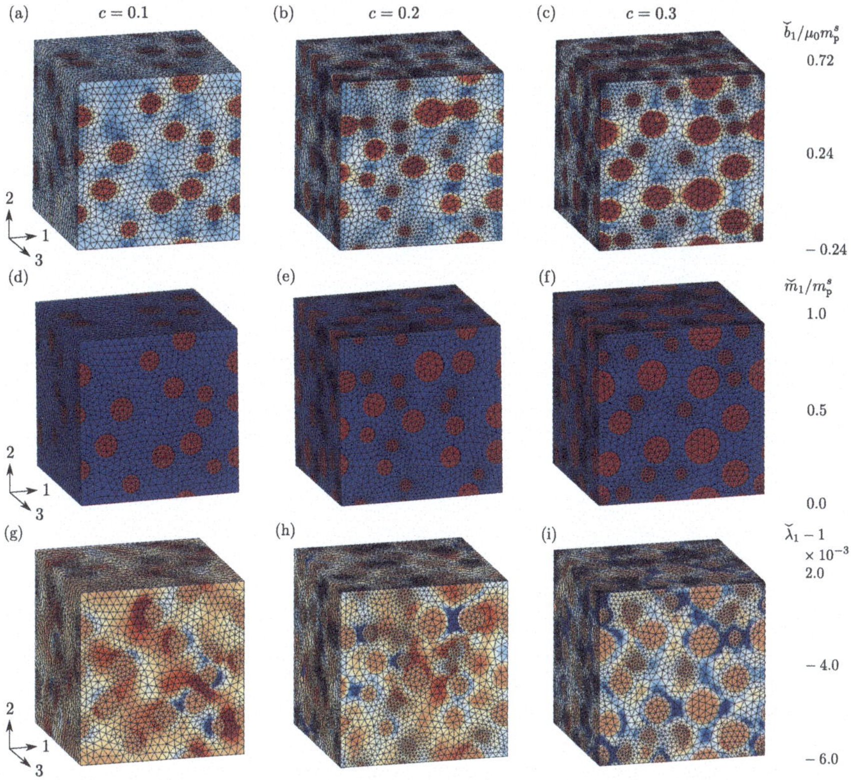

Fig. 8 Contours of the numerically computed **a–c** normalized microscopic $\breve{b}_1$, **d–f** normalized local $\breve{m}_1$ and **g–i** nominal mechanical strain $\breve{\lambda}_1 - 1$ in the RVEs *after* the first half cycle of loading along $\mathbf{e}_1$, under which the effective responses are shown in Fig. 7. Three different, RVEs having (**a, d, g**) $c = 0.1$, (**b, e, h**) 0.2 and (**c, f, i**) 0.3 are shown. (Taken from Mukherjee et al. (2021))

tive response is the closest to the corresponding average shown by the firm lines in Fig. 7b, c.

Then, the proposed **F**-**H** and **F**-**B** model responses are compared with the full-field numerical homogenization response in Fig. 9 for $c = 0.1,\ 0.2$ and 0.3. Here we consider a representative matrix shear modulus $G_{\mathrm{m}} = 0.5$ MPa, which corresponds to the shear modulus of the PDMS elastomer. Notice in Fig. 9a that the saturation magnetization of the h-MREs increases in an almost linear fashion with c, which resembles closely to the s-MREs responses (Lefèvre et al., 2017; Danas, 2017; Mukherjee et al., 2020). The coercivity b^c of the composite, however, undergoes very little change with the increase in c. Nonetheless, the effective susceptibilities χ^e and χ^r also increase with c, which can be observed clearly from Fig. 9a. Overall, the model predictions for the magnetization in the h-MRE match perfectly with the numerically computed effective response.

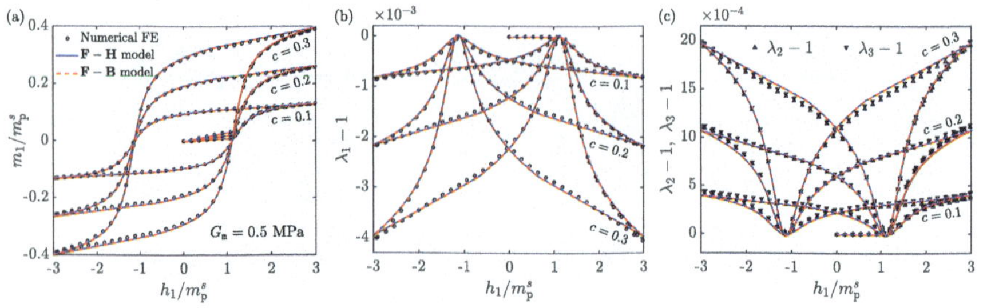

Fig. 9 Comparison of the effective **a** magnetization, **b** parallel and **c** transverse magnetostriction responses from the **F-H** and **F-B** models with the numerical homogenization estimates for a matrix shear modulus $G_{\mathrm{m}} = 0.5$ MPa. The h-MRE is subjected to an uniaxial cyclic h-field of magnitude $|h_1| = 3m_{\mathrm{p}}^s$ and the results for three different particle volume fractions of $c = 0.1,\ 0.2$ and 0.3 are shown. The average of the effective responses computed from five realizations of a RVE are shown without the fluctuation patches for the magnetostrictions. (Taken from Mukherjee and Danas (2022))

The local magnetostriction responses, on the other hand, exhibit a butterfly-shaped hysteresis loop with the applied cyclic magnetic field (see Fig. 9b, c). This response is essentially controlled by the coupling parameter β, which is calibrated from the numerically computed magnetostriction responses and has been provided in (64). Being the closed form complementary energy density, the **F-B** model does not need any further calibration. Thus, the same β parameter is used for the **F-B** model, yielding excellent match with the numerical homogenization response.

In this regard, we find that the model is capable of reproducing extremely well the effective magnetic response of the h-MRE for several volume fractions. As a result of this excellent agreement, the effective magnetostriction is also well reproduced by only a single calibration constant since β is a constant for a given volume fraction c.

In addition, we note that as the volume fraction of the particles decreases, the magnetization tends to saturate faster. By contrast, the switching point controlled by the magnetic coercivity b^c of the composite seems to be almost insensitive to the particle volume fraction, which justifies the proposition (69). We further note that the calibrated β parameter in (64) is also found to predict the effective magnetostriction responses considerably well for all $G_{\mathrm{m}} \geq 0.2$ MPa. Some representative computations to probe this predicting capability of the model have been carried out. These results are not shown here for brevity.

7.1.2 Effect of b_{p}^c and χ_{p}^e

In this section, we explore theoretically the response of an h-MRE material for different values of b_{p}^c and χ_{p}^e. We consider a shear modulus for the matrix phase $G_{\mathrm{m}} = 0.3$ MPa along with the magnetic particle parameters shown in Table 3 and the coupling parameter as in (64). Moreover, we use the inverse Langevin saturation

Table 3 Magnetic properties of the CIP particles in h-MRE formulation in Sect. 4

χ_{p}^{e}	χ_{p}^{r}	$\mu_0 m_{\mathrm{p}}^{s}$ (T)	b_{p}^{c} (T)	μ_0 (μN · A^2)
(0, 0.2, 0.4	30.0	2.5	0.05, 0.5, 1	$4\pi 10^{-1}$

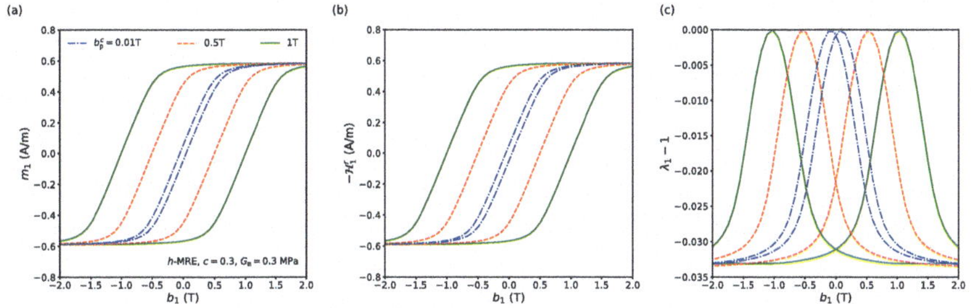

Fig. 10 Comparison of the **a** magnetization m_1, **b** (minus) internal variable $-\mathcal{H}_1^r$ and **c** magnetostriction response $\lambda_1 - 1$ as predicted by the h-MRE model for three values of $b_{\mathrm{p}}^{c} = 0.05, 0.5, 1$T, $\chi_{\mathrm{p}}^{e} = 0$, particle volume fraction $c = 0.3$ and matrix shear modulus $G_{\mathrm{m}} = 0.3$MPa

function defined in the last row of Table 1. We further control the **b** field and we impose overall

$$S_{ij}^{\mathrm{mech}} = 0, \ \forall i, j = 1, 2, 3, \qquad F_{ij} = 0, \ \ \forall i \neq j, \qquad \mathbf{b}^{\mathrm{app}} = b_1 \mathbf{e}_1. \tag{133}$$

The choice of these parameters in the limit of vanishing dissipation, i.e., $b_{\mathrm{p}}^{c} \to 0$ will allow in the next section to probe the h-MRE model response against the FE results for purely energetic s-MREs with carbonyl iron particle (CIP) inclusions, obtained in Mukherjee et al. (2020).

Specifically, Fig. 10 shows the prediction of the h-MRE models (recall that both versions **F-H** and **F-B** are equivalent) for three values of $b_{\mathrm{p}}^{c} = 0.05, 0.5, 1$T and $\chi_{\mathrm{p}}^{e} = 0$. It is plain from those graphs, that as $b_{\mathrm{p}}^{c} \to 0$ dissipation reduces to zero for all variables shown. Perhaps more interestingly, the internal variable $\boldsymbol{\mathcal{H}}^r$ becomes anhysteretic too. In fact, we have that $\boldsymbol{\mathcal{H}}^r = -\mathbf{m}$ when $\chi_{\mathrm{p}}^{e} = 0$. More importantly, it takes non-zero values in the limit $b_{\mathrm{p}}^{c} \to 0$.

When χ_{p}^{e} is not zero, as shown in Fig. 11, $\boldsymbol{\mathcal{H}}^r$ is directly linked to **m** but is not equal to $-\mathbf{m}$. Moreover, in the same figure, we observe that the parameter χ_{p}^{e} controls the unloading slope of the h-MRE. It was shown in Mukherjee and Danas (2019) that NdFeB powders have values of χ_{p}^{e} that range between $0.01 - 0.2$. This of course has important implications on the corresponding magnetization value upon complete removal of the external magnetic field as well as the corresponding magnetostriction which can reach much higher values when χ_{p}^{e} is large.

Remark 7.1 It is interesting to remark at this point that in essence, the magnetization variable may be seen as some form of *an internal variable* in the general dissipative model and not an independent one, as it is usually assumed in the literature (Brown,

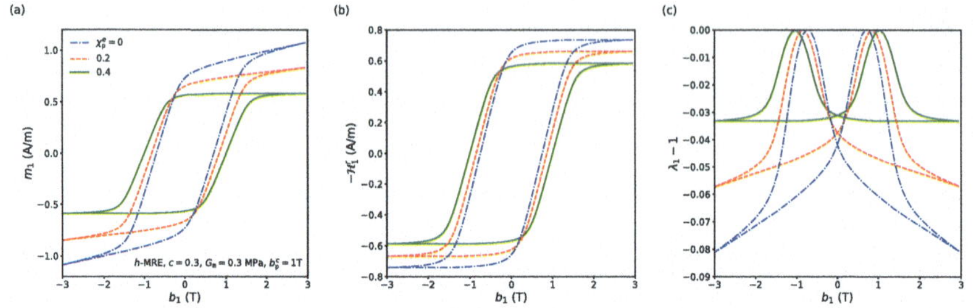

Fig. 11 Comparison of the **a** magnetization m_1, **b** (minus) internal variable $-\mathcal{H}_1^r$ and **c** magnetostriction response $\lambda_1 - 1$ as predicted by the h-MRE model for three values of $\chi_p^e = 0., 0.2, 0.4$, coercivity $b_p^c = 1$T, particle volume fraction $c = 0.3$ and matrix shear modulus $G_m = 0.3$MPa

1966; James & Kinderlehrer, 1993; Kankanala & Triantafyllidis, 2004; Danas et al., 2012). The reason is that **m** is directly related to the internal variable $\boldsymbol{\mathcal{H}}^r$ introduced originally in Mukherjee et al. (2021). This interpretation is in fact consistent with the fact that no boundary conditions can be imposed on **m** and thus no differential constraints. By contrast, an internal variable serves exactly that purpose, i.e., in addition to describe dissipation in the present context, it may be used to provide a measure of the internal state of the material similar to plastic strain in elasto-plasticity or polarization in electro-elasticity. In fact, the mechanism itself of magnetic domain motion inducing an internal state of magnetization or polarization may be thought in similar terms as the dislocation motion causes an internal state of plasticity. Another example of such a variable is the stress polarization in Hashin-Shtrikman estimates and again in that setting the stress polarization can exhibit jumps along interfaces or boundaries. This makes these polarization/magnetization variables powerful quantities to establish sometimes analytical approximate results. Nevertheless, they are not able to describe the material state entirely since either **B** or **H** still needs to be used so that actual boundary conditions can be imposed in a BVP. Finally, when only the energetic response of the material is analyzed the use of magnetization is in the general sense unnecessary.

7.1.3 The Limit of Zero Dissipation

Following the previous analysis, we now set $b_c = 10^{-6}$ and χ_p^e, keeping the remaining parameters presented in Table 3 and assess the h-MRE models by comparison with the FE results of (Mukherjee et al., 2020), which correspond to purely energetic s-MRE simulations. We plot the numerical FE response along with the model magnetization and magnetostriction responses in Fig. 12a–c, respectively. Besides the excellent agreement between the numerical homogenization computations and the model predictions, we observe two key differences between the s- and h-MREs by comparing Figs. 9 and 12. Firstly, and the obvious is the absence of hysteresis in the s-MREs. Secondly, s-MREs tend to saturate at a lower applied h_1, while h-MREs

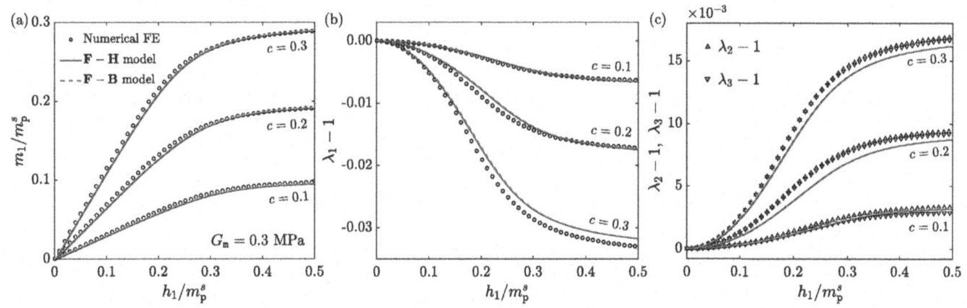

Fig. 12 Comparison of the **a** magnetization, **b** parallel and **c** transverse magnetostriction responses from the **F**-**H** and **F**-**B** models with the numerical homogenization estimates. The s-MRE is subjected to an uniaxial cyclic h-field of magnitude $|h_1| = 0.5m_p^s$ and the results for three different particle volume fractions of $c = 0.1,\ 0.2$ and 0.3 are shown. (Taken from Mukherjee and Danas (2022))

saturate very slowly to a constant magnetization. This is in agreement with the corresponding s-MRE experiments of Danas et al. (2012) as well as those presented in Mukherjee and Danas (2019) for hard magnets.

7.2 *Magnetization Independent of Stretching in MREs*

7.2.1 Uniaxial Tension Perpendicular to Pre-magnetization

In this section, we apply a purely mechanical uniaxial tension loading along the $\mathbf{e}_2$ direction, which is perpendicular to the direction $\mathbf{e}_1$ of the pre-magnetization of the h-MRE, while the applied magnetic field is kept identically zero, i.e.,

$$S_{22}^{\mathrm{mech}} \neq 0, \quad S_{11}^{\mathrm{mech}} = S_{33}^{\mathrm{mech}} = 0, \quad F_{ij} = 0, \ \ \forall i \neq j, \quad \mathbf{h}^{\mathrm{app}} = \mathbf{0}.$$

As shown in the inset of Fig. 13a, we consider half a cycle, described by a linear increase of S_{22}^{mech} from 0 to G_{m} and subsequent decrease to 0 (note that the rate of loading is inconsequential since the models under study are rate-independent). Moreover, we show results for three shear moduli, $G_{\mathrm{m}} = 0.3, 0.5, 1.0$.

Figure 13a–c show the mechanical stretch λ_2 and the magnetizations along the $\mathbf{e}_1$ and $\mathbf{e}_2$ directions for the numerical RVE and the analytical model. The corresponding deformed RVEs are depicted in Fig. 13d–f. It is noted that the numerically computed effective stretch $\lambda_2 - \lambda_2^0$ (with λ_2^0 denoting the initial remanent stretch due to the pre-magnetization) *does fluctuate* with the different RVE realizations. Nevertheless, as shown in Fig. 7, the magnitude of such realization-dependent scatter in the strain remain less than 5×10^{-4}, which is considerably smaller than the magnitude of the stretch ($\sim$0.25), shown in Fig. 13a. Thus, the numerical computations with the monodisperse RVEs lead to excellent estimates under purely mechanical loading conditions, while at the same time, they keep the computational expense considerably

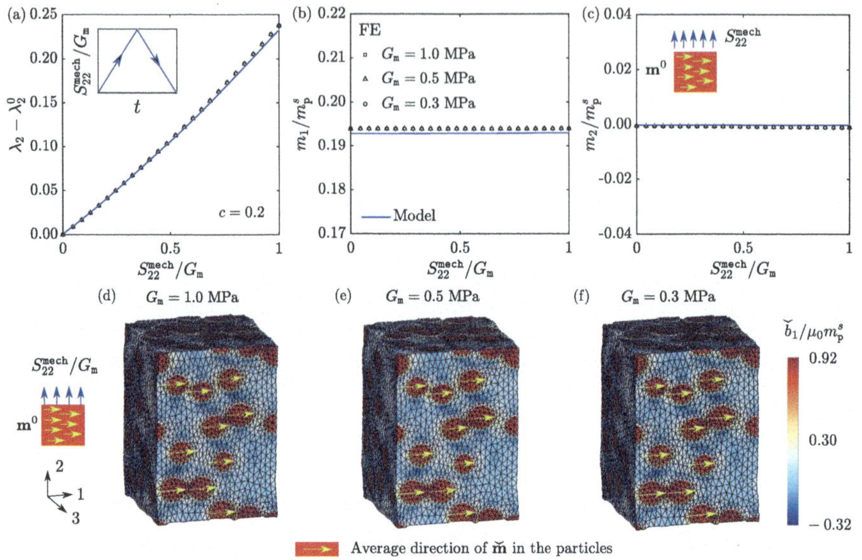

Fig. 13 Evolutions of **a** mechanical stretch $\lambda_2 - \lambda_2^0$ (with λ_2^0 denoting the initial remanent stretch due to the pre-magnetization) and magnetizations along **b** $\mathbf{e}_1$ and **c** $\mathbf{e}_2$ under applied uniaxial tensile stress S_{22}^{mech}, whose temporal evolution is shown in the inset of (**a**). The inset of (**c**) shows a schematic of the h-MRE with the direction of pre-magnetization $\mathbf{m}^0$ and the applied uniaxial tension. Contours of the $\breve{\mathbf{b}}$ field in the deformed RVE under applied $S_{22}^{\text{mech}}/G_{\text{m}} = 1$ for $G_{\text{m}} =$ **d** 1.0, **e** 0.5 and **f** 0.3 MPa. The arrows on the particles show the average direction of $\breve{\mathbf{m}}$ in them. (Taken from Mukherjee et al. (2021))

low. This observation has already been done in the context of s-MREs by Danas (2017), where the mechanical and magnetic response was found to converge for RVEs of considerably lower number of particles, whereas the pure magnetostriction (i.e. for zero overall applied mechanical load) required substantially larger RVE sizes with more particles.

In particular, we observe excellent agreement between the numerical homogenization results and the model estimates in all cases shown in Fig. 13a–c, namely, the principle stretch λ_2 and the effective magnetization responses along $\mathbf{e}_1$ and $\mathbf{e}_2$. The model predictions for the transverse stretches λ_2 and λ_3 also match perfectly the numerically computed responses (not shown explicitly here). It is noted further that all results shown here are independent of the matrix shear modulus upon the normalization $S_{22}^{\text{mech}}/G_{\text{m}}$. This is a particular feature of the Neo-Hookean model used for the mechanical description of the matrix phase and simply implies that the overall response of the h-MRE is also of a Neo-Hookean type at least to a very good approximation (see relevant discussion in Lopez-Pamies et al. (2013)).

Remark 7.2 Finally, we close the discussion of Fig. 13 with an important observation, that of the *stretch-independence* of the current effective remanent magnetization $\mathbf{m}^0$, observed in Fig. 13b, c, as predicted by the model and confirmed by the RVE simulations. In simple words, we find that the current remanent magnetization remains unaffected by the stressing (or stretching) of the solid. As a result, the mechanical cyclic loading of a pre-magnetized h-MRE does not lead to dissipation. This does not mean that the local magnetization does not change via corresponding particle rearrangement. On the contrary, particles rearrange due to the finite straining. Nonetheless, this does not affect the average current magnetization amplitude of the RVE, which is an important feature that needs to be reproduced both by phenomenological top-down as well as homogenization bottom-up models. The ability of the present model to recover this feature is linked to the definition of the internal variable $\mathcal{H}^r$ in the intermediate stretch-free configuration as discussed in Sect. 4.1 and the corresponding choice of invariants and coupled energy proposed. In turn, this feature has also been observed in the context of s-MREs experimentally (Danas et al., 2012), numerically (Mukherjee et al., 2020) and theoretically (Lefèvre et al., 2017) via an independence of the current magnetization response to pre-stressing. This feature is linked also to the underlying (quasi-)incompressibility of the materials under study and should be taken into account in the modeling of MREs in general. We also note that the same response is observed if the tension is parallel to the pre-magnetization direction (not shown here).

Remark 7.3 We further remark that this observation of the stretch independence of magnetization amplitude in incompressible MREs has been very recently confirmed by Yan et al. (2023) (but see also Yan et al. (2021a)) experimentally by pressurizing (and thus pre-stretching) thin plates. These authors have then proposed a modification of the original Zhao et al. (2019) theory. In particular, they have proposed that the current magnetization $\mathbf{m}$ is evaluated in terms of the pre-magnetization (reference) state $\mathbf{m}^0$ as $\mathbf{m} = \mathbf{R}\mathbf{m}^0$, instead of $\mathbf{m} = \mathbf{F}\mathbf{m}^0$. Given the direct connection between $\mathcal{H}^r$ and $\mathbf{m}$ shown in Fig. 10, this is equivalent to the present proposition of $\mathcal{H}^r$ being stretch-independent and defined in the intermediate configuration (see Fig. 2).

To go a step further, it is rather straightforward to consider a thought example, where a saturated pre-magnetized h-MRE along direction 1 (i.e. $m_1^0 = m^s$) is subjected to a stretch along direction 1, i.e., $F_{11} > 1$. By using the original proposition, $\mathbf{m} = \mathbf{F}\mathbf{m}^0$, it would simply imply that the current magnetization $m_1 = F_{11} m_1^0 > m^s$ is (much) larger than the saturation magnetization of the h-MRE, which is practically and theoretically impossible. On the contrary, $\mathbf{m} = \mathbf{R}\mathbf{m}^0$ would simply lead to $m_1 = m_1^0 = m^s$, a result that is theoretically consistent and as was shown numerically and experimentally is the case for incompressible s- and h-MREs.

7.2.2 Simple Shear Parallel to Pre-magnetization

We apply a purely mechanical simple shear stress S_{12}^{mech} loading. The corresponding traction vector is parallel to the pre-magnetization direction $\mathbf{e}_1$ of the h-MRE, while the applied magnetic field is kept identically zero during this step, i.e.,

$$S_{12}^{\text{mech}} \neq 0, \quad S_{11}^{\text{mech}} = S_{22}^{\text{mech}} = S_{33}^{\text{mech}} = 0, \quad \mathbf{h}^{\text{app}} = \mathbf{0},$$

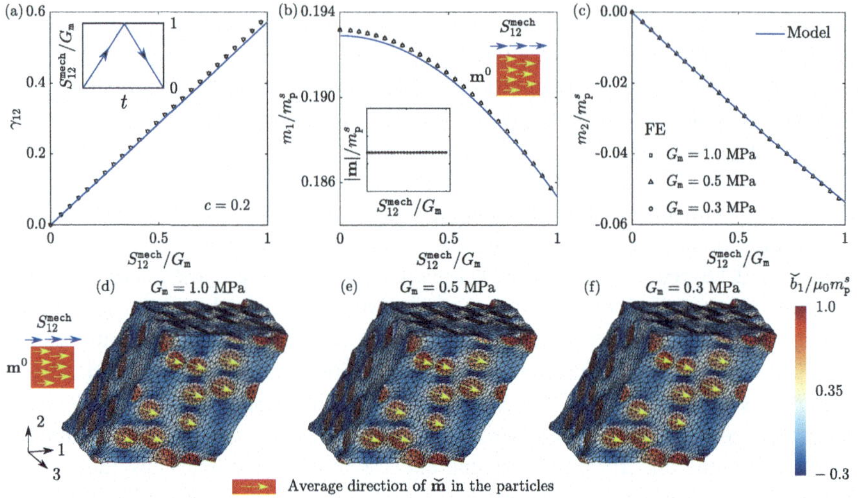

Fig. 14 Evolutions of **a** shear strain γ_{12} and magnetizations along **b** $\mathbf{e}_1$ and **c** $\mathbf{e}_2$ under applied simple shear stress S_{12}^{mech}, whose loading path is shown in the inset of (**a**). A schematic of the h-MRE with the direction of pre-magnetization $\mathbf{m}^0$ and the applied shear stress is shown in the inset of (**b**). Contours of the $\breve{\mathbf{b}}$ field in the deformed RVE under applied $S_{12}^{\mathrm{mech}}/G_{\mathrm{m}} = 1$ for $G_{\mathrm{m}} =$ **d** 1.0, **e** 0.5 and **f** 0.3 MPa. The arrows on the particles show the average direction of $\breve{\mathbf{m}}$ in them. (Taken from Mukherjee et al. (2021))

together with $F_{21} = F_{13} = F_{31} = F_{23} = F_{32} = 0$. Furthermore, the evolution for the applied S_{12}^{mech} is shown in the inset of Fig. 14a.

In Fig. 14a–c, we observe an excellent agreement between the model predictions and the numerical homogenization results for the effective shear strain $F_{12} = \gamma_{12}$ as well as for the effective magnetizations along $\mathbf{e}_1$ and $\mathbf{e}_2$, respectively. All results, shown in the context of this figures, are independent of the matrix shear modulus upon the normalization $S_{12}^{\mathrm{mech}}/G_{\mathrm{m}}$.

Again, we observe that despite the significant shearing strains and particle rearrangements, the amplitude of the current effective magnetization $\mathbf{m}$ remains unaffected (see inset of Fig. 14b). Instead, the orientation of the magnetization vector significantly changes with the applied shearing, as revealed by the change of the individual components m_1/m_{p}^s and m_2/m_{p}^s in Fig. 14b, c, respectively. Interestingly, this rotation remains (almost) identical to the macroscopic (average) rotation of the RVE induced by the shearing. Therefore, the affine rotation-based model presented in this work (as well as that of Yan et al. (2023)) predicts the evolution of $\mathbf{m}$ in this case accurately.

Figures 14d–f show three deformed RVEs at $S_{12}^{\mathrm{mech}}/G_{\mathrm{m}} = 1$ for the three G_{m} under consideration. It is noted that the RVE deformations and the local $\breve{b}_1$ fields remain identical for all three G_{m} under consideration. The only key feature to note here is the uniform distribution of the particle rotations, which, in turn, rotates the local (microscopic) and therefore the global (effective) magnetization vectors.

7.3 NdFeB-Based h-MRE Versus CIP-Based s-MRE Response

In this section, we provide a useful set of results that serve to discuss practical differences between actual h-MREs with NdFeB particles and s-MREs with CIP particles. In the very recent literature, an impressive amount of studies has focused mainly on h-MREs subjected to small magnetic fields at their fully pre-magnetized state. While this state exploits mainly magnetic torques in slender objects, it is only a small fraction of the response of the more general class of MRE materials and potential application (see for instance Moreno-Mateos et al. (2022), Garcia-Gonzalez et al. (2023)). We use for simplicity the same inverse Langevin saturation function defined in the last row of Table 1 for both h- and s-MRE results shown next. This changes only slightly the transition response of h-MREs but not the initial and final saturation response. In order to cover a variety of effects, we will include in the following also pre-stress effects. In the work of Danas et al. (2012) related to CIP-filled s-MREs, it was shown that the magnetization response is almost entirely insensitive to the pre-stress. In contrast, a strong effect of the pre-applied mechanical load was observed for the magnetostriction. In the following, we will consider various combinations of the pre-stresses. The magnetic field is always applied along direction 1, i.e., $\mathbf{b} = b_1 \mathbf{e}_1$, and the matrix phase has a shear modulus $G_{\mathrm{m}} = 0.05\mathrm{MPa}$.

In this regard, we analyze first in Fig. 15 the effect of uniaxial pre-stressing parallel to the applied magnetic field, such that

$$S_{11}^{\mathtt{mech}} = \{-1, 0, 1\}, \quad S_{22}^{\mathtt{mech}} = S_{33}^{\mathtt{mech}} = 0, \quad F_{ij} = 0, \quad \forall i \neq j, \quad \mathbf{b} = b_1 \mathbf{e}_1.$$

We note first that the pre-stress has no effect on the magnetization response in Fig. 15a. This is consistent with the early work of Danas et al. (2012) on s-MREs, while we observe that the same feature is true for the h-MREs too. Moreover, the CIP-based s-MRE has a much higher saturation magnetization and initial permeability than that for the NdFeB. This allows for the s-MRE material to reach much higher magnetostrictive strains and at smaller magnetic fields as clearly shown in Fig. 15b. In turn, the h-MRE exhibits important dissipative effects and much lower magnetostrictive strains. Similarly, as a direct consequence of this feature, the s-MRE is expected to exhibit a stronger "magnetorheological" effect (i.e., increase in the apparent shear or Young's modulus) upon the application of a magnetic field (Diguet et al., 2021). In turn, the h-MRE retains a permanent magnetization in the absence of an applied magnetic field and thus is more relevant for torque-based or permanent magnetic applications. We also observe that tensile pre-stresses increase the resulting amplitude of the magnetostriction (which remains always negative in this homogenization analysis), while compressive ones lead to a decrease of $|\Delta\lambda_1|$. This observation is the same for both MREs considered here.

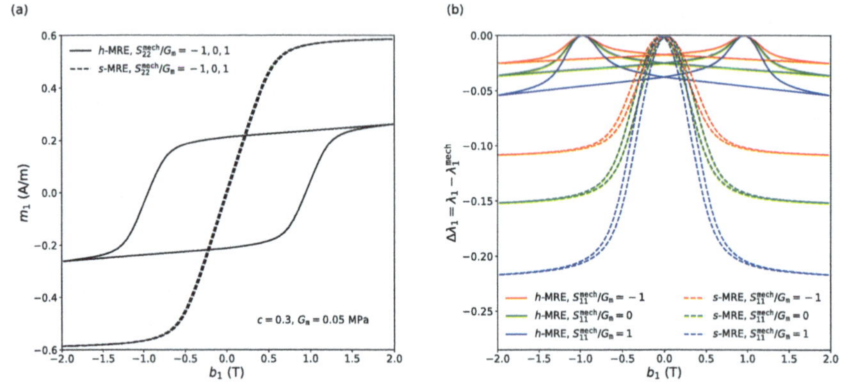

Fig. 15 Comparison of the **a** magnetization and **b** magnetostriction response using the material properties defined in Table 2 for the h-MRE with NdFeB particles and Table 3 for the s-MRE with CIP particles. Three uniaxial prestress values $S_{11}^{\rm mech}/G_{\rm m} = \{-1,\ 0,\ 1\}$ are considered along the applied magnetic field. We set the particle volume fraction to $c = 0.3$ and matrix shear modulus $G_{\rm m} = 0.05$MPa. Magnetostriction in this figure is defined as the magnetically induced strain minus the mechanical strain induced by the mechanical prestress, i.e. $\Delta\lambda_1 = \lambda_1 - \lambda_1^{\rm mech}$

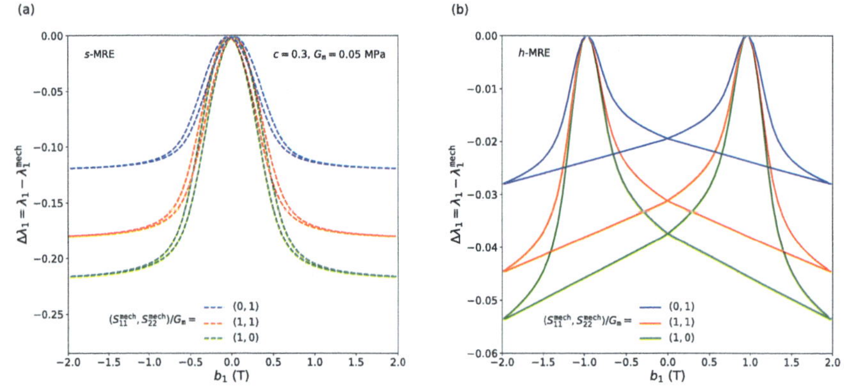

Fig. 16 Comparison of the magnetostriction response using the material properties defined **a** in Table 3 for the s-MRE with CIP particles and **b** in Table 2 for the h-MRE with NdFeB particles. Three prestress sets of values $(S_{11}^{\rm mech}, S_{22}^{\rm mech})/G_{\rm m} = (1, 0)|(1, 1)|(0, 1)$ are considered. We set the particle volume fraction to $c = 0.3$ and matrix shear modulus $G_{\rm m} = 0.05$ MPa. Magnetostriction in this figure is defined as the magnetically induced strain minus the mechanical strain induced by the mechanical prestress, i.e. $\Delta\lambda_1 = \lambda_1 - \lambda_1^0$

We close the section by considering in Fig. 16, magnetostriction curves for (a) the CIP-based s-MRE and (b) NdFeB h-MRE for three sets of triaxial pre-stress values

$$(S_{11}^{\rm mech}, S_{22}^{\rm mech})/G_{\rm m} = (0, 1)|(1, 1)|(1, 0), \qquad S_{22}^{\rm mech}/G_{\rm m} = S_{33}^{\rm mech}/G_{\rm m},$$

$$F_{ij} = 0, \quad \forall i \neq j, \qquad \mathbf{b} = b_1\mathbf{e}_1.$$

Specifically, Fig. 16a,b shows that uniaxial tension pre-stressing along the magnetic field leads to higher amplitude of magnetostrictive strains than tension along the perpendicular direction. A hydrostatic tension pre-stress leads to a magnetostriction $\Delta\lambda_1$ that lies in-between the two. This figure reveals clearly the strong effect of pre-stress upon the magnetostriction and by extension to the magnetorheological effect, which would be present if instead of pre-stressing a pre-straining was applied. Again, the magnetization is not affected in this case by the pre-stresses and is the same with that shown in Fig. 15.

7.4 *Energetic s-MRE Models Versus h-MRE Models with Zero Dissipation*

In Sect. 4, we have provided a unified modeling framework for isotropic, incompressible hard and soft MREs. The latter is obtained by considering the limit of $b_{\mathrm{p}}^{c} \to 0$. However, in practice this limit is rather difficult to consider analytically since in that limit $\mathcal{H}^{r}$ becomes a nonlinear function of **H** or **B**, depending on which formulation one uses.

An alternative purely energetic approach, which however, was introduced earlier than the full dissipative one, has been discussed in Sect. 5 and originally presented in Mukherjee et al. (2020). This approach is not dual as already discussed in the aforementioned section but, nonetheless, provides both an **F**-**H** and **F**-**B** model that are close to each other. In the original work of Mukherjee et al. (2020), those models were calibrated against the analytical, implicit homogenization model of Lefèvre et al. (2017). As it was shown in that work, while the proposed models (all of them by construction) do well for CIP volume fractions of $c \leq 0.2$, they tend to underestimate the magnetostriction (but not the magnetization) response for $c > 0.25$ when compared with corresponding FE periodic results. In the present manuscript, we have introduced the new function (92) to fix this discrepancy.

In view of this, we provide in Figs. 17 and 18 a comparison between the results obtained previously in the case of vanishing dissipation by the h-MRE model with $b_{\mathrm{p}}^{c} \to 0$ and the s-MRE models presented in Sect. 5. For the s-MRE models we use the parameters provided in Table 4.

In particular, we observe that for both matrix moduli $G_{\mathrm{m}} = 0.05,\ 0.3$ MPa considered here as an example, all models lie fairly close to each other. Moreover, we observe no dependence of the magnetization response on the moduli of the matrix phase. The h-MRE model is considered as the reference case given its excellent agreement with the FE simulations discussed in the previous section. In this view then, the **F**-**H** (**F**-**B**) s-MRE model tends to slightly underestimate (overestimate) the magnetostriction amplitude at large volume fractions (e.g., $c = 0.3$). In turn, the s-MRE models in general underestimate slightly the magnetization response. For volume fractions $c \leq 0.2$, the agreement between all models is excellent. In this regard, we conclude in this study that any of the above models may be used to

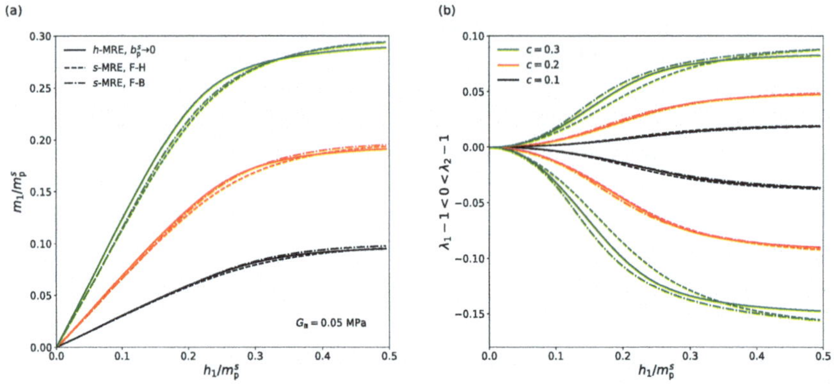

Fig. 17 Comparison of the **a** magnetization and **b** magnetostriction response as predicted by the s-MRE models discussed in Sect. 5 and the reference h-MRE models at the limit of $b_p^c \to 0$ for three particle volume fractions $c = 0.1,\ 0.2$ and 0.3 and matrix shear modulus $G_m = 0.05$MPa

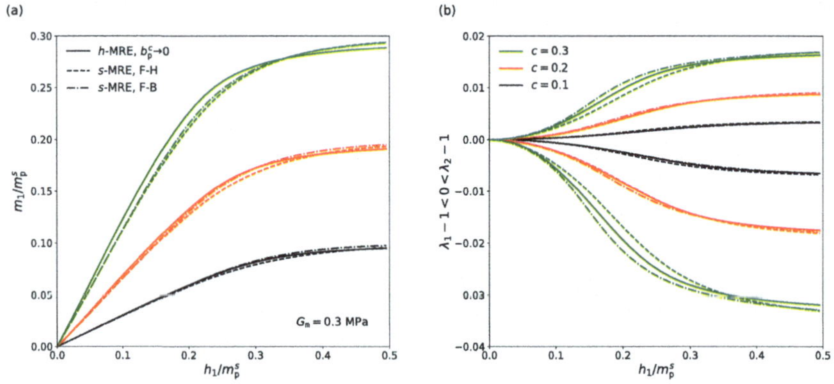

Fig. 18 Comparison of the **a** magnetization and **b** magnetostriction response as predicted by the s-MRE models discussed in Sect. 5 and the reference h-MRE models at the limit of $b_p^c \to 0$ for three particle volume fractions $c = 0.1,\ 0.2$ and 0.3 and matrix shear modulus $G_m = 0.3$MPa

Table 4 Magnetic properties of the CIP particles in s-MRE formulation in Sect. 5

χ_p	$\mu_0 m_p^s$ (T)	μ_0 (μN · A^2)
30.0	2.5	$4\pi 10^{-1}$

model s-MREs depending on the problem at hand and convenience. Obviously, the purely energetic s-MRE models are easier to implement since they do not require any definition of internal variables or incremental procedures (as described in the general case in Sect. 6).

8 Results: Numerical BVP Simulations

This section shows numerically computed boundary value problem (BVP) solutions for s-MREs as well as uniformly and non-uniformly pre-magnetized h-MRE beams. The following results make use of the previously discussed models, which are numerically implemented in user-element Abaqus subroutines and consider the soft and hard particle magnetization parameters as in Tables 4 and 2, respectively. The particle volume fraction c and the loading conditions for pre-magnetization and/or actuation steps are discussed under specific subsections depending on the examples analyzed. Moreover, the specific Dirichlet boundary conditions on $\mathbf{u}_T$, φ_T and α_T, i.e, the case-specific versions of the admissible sets $\mathcal{U}$ in (108), $\mathcal{G}$ in (109) and $\mathcal{B}$ in (112), respectively, are detailed in the following.

8.1 Generic Numerical BVP Setting

The discretization of the scalar and vector potential-based variational principles were discussed in a fairly general setting so far in Sect. 6. We now specify a geometry for the numerical BVP of interest. Although the MREs are finding applications in a wide variety of engineering devices, such as in actuators most of the recent attention is in fabrication and testing of slender structures, which find applications in soft robotic devices (Kim et al., 2018; Ren et al., 2019) as well as in thin membranes or films Psarra et al. (2017), Psarra et al. (2019), Moreno-Mateos et al. (2022). This includes spatially uniformly and non-uniformly pre-magnetized beams, functionally graded beams with a distribution of the particle volume fractions c, as well as films resting substrates.

In particular, we consider a representative (but otherwise generic) two-dimensional, plane-strain analysis of the bending of pre-magnetized slender beams. We emphasize in this context, that the magnetic fields are applied via the fixed electromagnet poles far away from the MRE (not modeled explicitly here), both during the pre-magnetization and actuation. Thus, it is necessary to embed the MREs in a surrounding air. Moreover, since the magnetic fields are applied far away (or at a given distance), the air domain is considered to be substantially larger than the MRE. In particular, let us consider the air domain length $L > \ell$ (e.g. ten times larger) to ensure that the MRE deflection is sufficiently far from the boundary of the air $\partial\mathcal{V}^{\text{Top}}_{\text{Air}} \cup \partial\mathcal{V}^{\text{Right}}_{\text{Air}} \cup \partial\mathcal{V}^{\text{Bottom}}_{\text{Air}} \cup \partial\mathcal{V}^{\text{Left}}_{\text{Air}}$ (see Fig. 19a).

As shown in Fig. 19b, the slender MRE beam of length ℓ and width w has a common interface $\partial\mathcal{V}^{\text{Top}}_{\text{MRE}} \cup \partial\mathcal{V}^{\text{Right}}_{\text{MRE}} \cup \partial\mathcal{V}^{\text{Bottom}}_{\text{MRE}} \cup \partial\mathcal{V}^{\text{Left}}_{\text{MRE}}$ with the surrounding air. The aspect ratio of the beam is hence defined via $r_{\text{asp}} = \ell/w$. Finally, the structured FE mesh used in the computations is shown in Fig. 19c. Throughout this paper we consider linear four-node quadrilateral isoparametric elements in the FE computations.

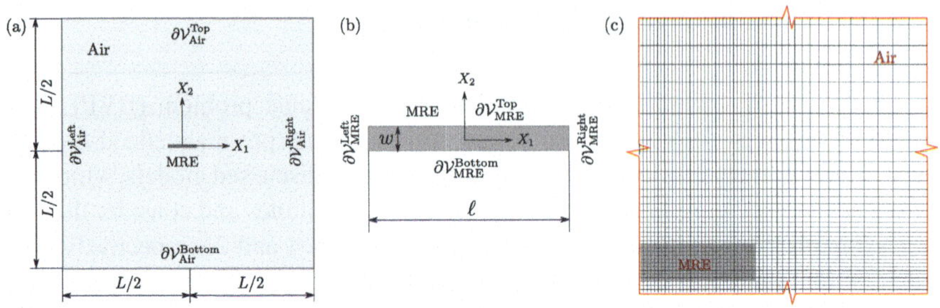

Fig. 19 **a** Diagram of the full BVP domain having MRE and the surrounding air. The air domain with an external boundary $\partial\mathcal{V}_{\text{Air}}^{\text{Top}} \cup \partial\mathcal{V}_{\text{Air}}^{\text{Right}} \cup \partial\mathcal{V}_{\text{Air}}^{\text{Bottom}} \cup \partial\mathcal{V}_{\text{Air}}^{\text{Left}}$ is considered to be a square of length L. The reference coordinate system $\mathbf{X}$ is considered to have origin at the center of the air domain. **b** Dimensions of the MRE having the interface $\partial\mathcal{V}_{\text{MRE}}^{\text{Top}} \cup \partial\mathcal{V}_{\text{MRE}}^{\text{Right}} \cup \partial\mathcal{V}_{\text{MRE}}^{\text{Bottom}} \cup \partial\mathcal{V}_{\text{MRE}}^{\text{Left}}$ with the surrounding air. The MRE length ℓ is considered to be $\ell = 0.1L$ and the aspect ratio of the MRE is defined as $r_{\text{asp}} = \ell/w$. **c** A part of the structured mesh considered in the calculations. Standard linear 4-node quadratic isoparametric elements are employed. (Taken from Mukherjee and Danas (2022))

8.2 *Treatment of Air*

The air surrounding the MRE has (nearly) zero mechanical stiffness, whereas the magnetic $\mathbf{b}$ and $\mathbf{h}$ fields in it are finite. Specifically, the former is related to the latter via $\mathbf{b} = \mu_0\mathbf{h}$ in the surrounding air. Dealing with a material of nearly zero mechanical stiffness in the present fully implicit, Lagrangian modeling framework leads to extreme mesh distortions at the corners of the MRE, eventually stopping the numerical simulation from converging.

Till this date, a number of methods have been implemented for dealing with the surrounding air in the magneto-active structures. The most straightforward way to model the air is to consider it a nearly incompressible or compressible hyperelastic solid having shear modulus of $\sim$1 Pa (Rambausek & Keip, 2018; Dorn et al. 2021). However, such an assumption may lead to an underestimation of the mechanical deformations of the MREs, specifically when undergoing large deformations or deflections. An alternative approach, namely, the method of constraining the motion of the air nodes surrounding the MRE is found to yield very accurate results of MRE deformations in air (Psarra et al., 2019; Mukherjee et al., 2021). In particular, the latter considers the air shear modulus to be zero but simultaneously applies linear constraints on all nodes in the air domain to make them move according to the deformation/deflection of the MRE boundary. Having said that, we also remark that the application of such linear constraints on the air nodes, where two or more (magnetic or non-magnetic) structures are interacting, may become difficult to implement properly so that numerical convergence is achieved. A quantitative comparison of the performance of different modeling approaches for the surrounding air is drawn in a recent paper (Rambausek et al., 2022). Two additional promising approaches, not discussed here, have been proposed only recently in Rambausek and Schöberl

(2023), where a proper treatment of the Maxwell stress at the interface between the magnetoelastic solid and the air allows to eliminate the spurious modes present in such problems and allow for very good convergence. Another potential solution to the problem could be the use of meshfree methods (see for instance Kumar et al. (2019)). Therein, it was shown that very large strains may be reached at soft regions of the domain in a straightforward manner. Nonetheless, those methods are not yet available in more general-use software packages and thus their use is less visited.

In this paper, we consider standalone MRE solids that are subjected to spatially uniform magnetic fields, as shown in Fig. 19a. Thus, we employ the air node constraining method to model the deformation in the domain $\mathcal{V}_{\text{Air}}$. In fact, the linear constraints on the displacement field $\mathbf{u}$ for all $\mathbf{X} \in \mathcal{V}_{\text{Air}}$ can be applied via directly augmenting the incremental variational principles (106) and (110) by a penalty energy instead of potential

$$W_{\texttt{penalty}}(\mathbf{u}) = \sum_{j=1}^{N_{\text{Air}}} \sum_{i=1}^{2} \frac{G_{\text{c}}}{2L_{\text{c}}\zeta} \left(\mathcal{C}_i^{(j)}\right)^2, \tag{134}$$

where N_{Air} is the number of air nodes, L_{c} is a reference length parameter usually considered to be equal to w, G_{c} is an arbitrary shear modulus that we consider to be identical to that of the matrix and ζ is the penalty parameter, which is set to 10^{-3}. Nevertheless, any value of ζ in the range 10^{-6}–10^{-3} ensures a proper imposition of the constraint not affecting the numerical convergence significantly. Given that those constraints are linear one has also the option to directly use the `*Equation` command in Abaqus. Such an approach has also been tested showing no differences with the penalty approach described here in two and three dimensions. Finally, the pointwise constraint $\mathcal{C}_i^{(j)}$ is defined as (Psarra et al., 2019)

$$\mathcal{C}_i^{(j)} \equiv \begin{cases} d_i^{(j)} u_i^{(j)}\Big|_{\partial\mathcal{V}_{\text{MRE}}} - u_i^{(j)}\Big|_{\mathcal{V}_{\text{Air}}} = 0, & \text{if } 0 < d_i^{(j)} \leq 1 \\ u_i^{(j)}\Big|_{\mathcal{V}_{\text{Air}}} = 0, & \text{otherwise,} \end{cases} \tag{135}$$

which ties the displacement of any node j in $\mathcal{V}_{\text{Air}}$ with that of its nearest node on the Air/MRE interface $\partial\mathcal{V}_{\text{MRE}}$. In practice, we construct a set of two-node elements comprising one node from $\mathcal{V}_{\text{Air}}$ and one from the set $\partial\mathcal{V}_{\text{MRE}}$ that has the least Euclidean distance from the former. Subsequently, we add the "force" and "stiffness" terms to the global force and stiffness matrices. Those terms emerge by considering first and second variations of the corresponding degrees-of-freedom involved the penalty energy (134).

The constraint "weight" function $d_i^{(j)}$ is defined in terms of the absolute distance difference between the X_i $(i = 1, 2)$ coordinates of the points in $\mathcal{V}_{\text{Air}}$ and on $\partial\mathcal{V}_{\text{MRE}}$, such that

$$d_i^{(j)} = 1 - \frac{\left| X_i^{(j)}\big|_{\mathcal{V}_{\text{Air}}} - X_i^{(j)}\big|_{\partial\mathcal{V}_{\text{MRE}}} \right|}{R_f L/2}, \qquad \text{with } i = 1, 2. \tag{136}$$

Here, $R_f \in (0.5, 1]$ is a fraction coefficient that is user-defined and serves to prescribe the range of the deformable air. The air nodes lying outside that region are simply fixed and are not allowed to displace. The penalty energy (134) may be added directly to the general variational principles (106) or (110) or simply impose the linear constraints (135) via an elimination technique (such as the `*Equation` command in Abaqus).

8.3 Magnetostriction and Magnetization Response of a Spherical s-MRE Specimen

In this section, we show qualitative results for the ideal problem of an s-MRE spherical specimen embedded in a large spherical air domain as shown in Fig. 20a. A remotely applied magnetic field induces magnetostrictive strains and magnetization inside the inclusion. This problem has been the focus of various studies starting from the seminal manuscript of Brown (1966), who proposed a solution in the small strain setting. Therein, he has shown that the presence of a magnetically-induced traction, i.e., the magnetic part of the Maxwell stress (see for instance Eq. (103)) leads to non-uniform mechanical fields in the magneto-elastic inclusion. Here, we recall the analysis carried out in Lefèvre et al. (2017, 2019) in the context of finite strains. This is only possible numerically.

We follow the approach of Lefèvre et al. (2017) wherein, for computational expediency, numerical solutions in the specimen and surrounding space—assumed to be air—are generated on a spatial domain of sufficiently large but finite extent, and not on $\mathbb{R}^3$ entirely. While full details of this approach can be found in Sect. 6 in (Lefèvre et al., 2017), it is appropriate to mention here that (*i*) the finite domain of computation is comprised of the spherical MRE specimen surrounded by an air-filled thick spherical shell subjected on its external surface to the affine boundary conditions $\mathbf{x} = \mathbf{X}$ and $\varphi = -\mathbf{H}_\infty \cdot \mathbf{X}$ (see Fig. 20a), (*ii*) the surrounding air is treated as a highly compressible magnetoelastic material with vanishingly small mechanical stiffness, and (*iii*) the numerical solutions are generated by means of a conforming axisymmetric 7-node hybrid triangular finite element discretization that leverages the axial symmetry of the problem around the direction, say $\mathbf{e}_3$, of the applied magnetic field $\mathbf{H}_\infty = H_\infty \mathbf{e}_3$.

By providing *pointwise* solutions for the deformation and magnetic fields in the MRE specimen and surrounding air, this approach also allows to extract *global* information about the deformation and magnetization of the specimen as would be done experimentally (Diguet et al., 2010; Diguet, 2010).

Figure 21 presents contour plots in the $\mathbf{e}_1$-$\mathbf{e}_3$ plane of the local component $F_{33}(\mathbf{X})$ of the deformation gradient and of the local component $m_3(\mathbf{x})$ of the magnetization over spherical specimens made of s-MREs containing $c = 0.222$ volume fraction of CIP particles. We use the same material properties for the CIP particles introduced in Table 4, while the matrix phase is taken with a shear modulus $G_{\mathrm{m}} = 50$kPA. The contours in Fig. 21a–c are shown over the undeformed configuration of the specimen

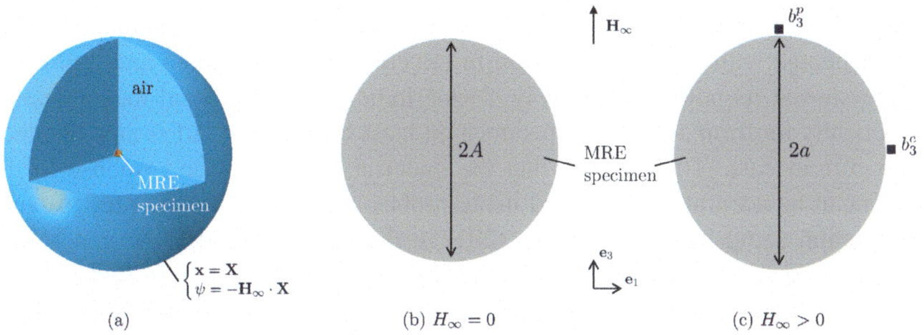

Fig. 20 **a** Schematic of the finite domain utilized to generate numerical solutions for the BVP of a s-MRE sphere embedded in air. The air domain is defined by a spherical shell of initial outer radius that is twenty times that of the MRE specimen. Schematics of a spherical MRE specimen of initial radius A in its **b** undeformed and **c** deformed configuration. (Taken from Lefèvre et al. (2019))

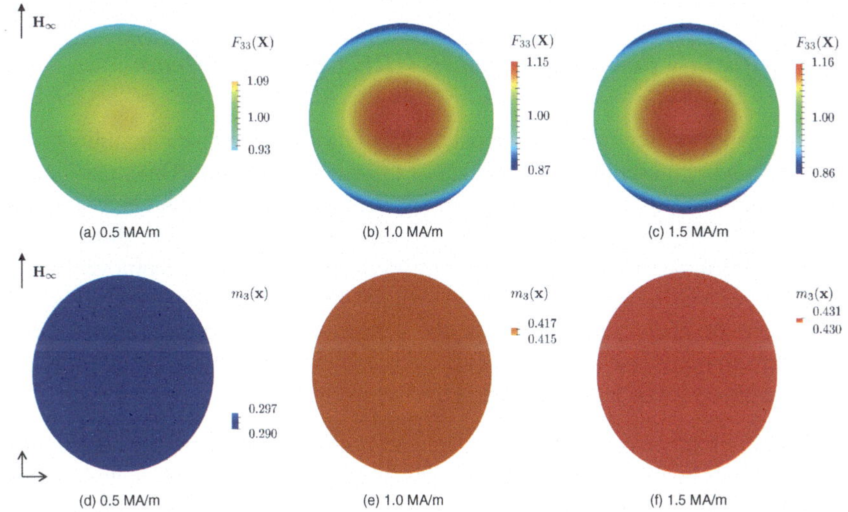

Fig. 21 Contour plots of **a–c** the component $F_{33}(\mathbf{X})$ of the deformation gradient over the *undeformed* configuration, and **d–e** the component $m_3(\mathbf{x})$ of the magnetization over the deformed configuration of a spherical specimen made of a s-MRE containing $c = 0.222$ volume fraction of iron particles. The contours correspond to the remotely applied magnetic field $\mathbf{H}_\infty = H_\infty \mathbf{e}_3$ with H_∞=0.5, 1.0, 1.5 MA/m. (Taken from Lefèvre et al. (2019))

as implied by the argument $\mathbf{X}$ of $F_{33}(\mathbf{X})$, while the contours in Fig. 21d–e are shown over the deformed configuration of the specimen as implied by the argument $\mathbf{x}$ of $m_3(\mathbf{x})$. Further, the contours correspond to the magnitudes $H_\infty = 0.5, 1.0, 1.5$ MA/m of the remotely applied magnetic field $\mathbf{H}_\infty$, and the color scale bars in each of them indicate the corresponding variation of the quantity of interest from its minimum to its maximum.

It is also clear from Fig. 21a–c that the local deformation gradient is highly heterogeneous, with regions in tension in the core of the specimen and regions in compression at its poles. In turn, Fig. 21d–e indicate that the local magnetization is practically uniform across the specimen, at least for the range of strains obtained in the current case. This implies that the material magnetization response of the s-MRE can be accurately measured using probes at its boundary (see for instance experimental setup in Bodelot et al. (2017)); the same is *not* true in general for cylindrical specimens (Bodelot et al., 2017; Lefèvre et al., 2017). In that case, all fields are highly heterogeneous. This implies that most experimental results available in the literature should be analyzed with extreme caution by theoreticians who attempt to propose material models, since the experimental measurements involve significant structural effects that sometimes are predominant over the corresponding material response.

8.4 Uniformly Pre-magnetized h-MRE Cantilever Beams

In the following two sections, we consider h-MREs with matrix shear modulus $G_{\rm m} = 0.187$ MPa, which resembles closely that of the moderately-soft PDMS elastomers (Kim et al., 2018; Zhao et al., 2019). Moreover, the matrix bulk modulus is considered to be $G'_{\rm m} = 500 G_{\rm m}$, which ensures a *nearly incompressible* material response.

We start with the simplest case of the uniformly pre-magnetized cantilever beams with the aspect ratios $r_{\rm asp} = 10$ and 17.5. Specifically, we simulate the experimental observations of Zhao et al. (2019) for the deflection of pre-magnetized h-MREs under uniform transverse actuation fields. To accomplish that, the loading is divided in to be two steps, which are detailed in the following.

- `Step-I`: First, we carry out the pre-magnetization along $\widehat{\mathbf{E}}_1$ by considering the air and MRE boundaries to be fixed. Thus, the Dirichlet boundary conditions on $\mathbf{u}$ and φ for the **F**-**H** model reads

$$\mathbf{u}_\tau = \mathbf{0}, \ \forall\, \mathbf{X} \in \partial\mathcal{V}_{\rm MRE}, \ \text{and} \ \mathbf{u}_\tau = \mathbf{0}, \ \forall\, \mathbf{X} \in \partial\mathcal{V}_{\rm Air} \tag{137}$$

$$\varphi_\tau = 0, \quad \forall\, \mathbf{X} \in \partial\mathcal{V}^{\rm Left}_{\rm Air}, \ \text{and} \ \varphi_\tau = -\frac{b^{\rm mag}_{1,\tau}}{\mu_0} L, \quad \forall\, \mathbf{X} \in \partial\mathcal{V}^{\rm Left}_{\rm Air}, \tag{138}$$

where $b^{\rm mag}_{1,\tau}$ is the magnetization field at time τ. In particular, $b^{\rm mag}_{1,\tau}$ is increased linearly in time up to 2 T followed by its decrease at the same rate to 0T. The rate of $b^{\rm mag}_{1,\tau}$ (and all the following applied fields) is inconsequential in the simulations since the material model is rate-independent. Similarly, the Dirichlet boundary condition on α for the **F**-**B** model reads

$$\alpha_\tau = 0, \quad \forall\, \mathbf{X} \in \partial\mathcal{V}^{\rm Bottom}_{\rm Air} \quad \text{and} \quad \alpha_\tau = b^{\rm mag}_{1,\tau} L, \quad \forall\, \mathbf{X} \in \partial\mathcal{V}^{\rm Top}_{\rm Air}, \tag{139}$$

while that on $\mathbf{u}_\tau$ remains identical to (137).

- `Step-II`: Next, we carry out the actuation step, where we apply a uniform field b_2^{actu} along $\widehat{\mathbf{E}}_2$, i.e., transverse to the centerline of the beam. The magnitude of b_2^{actu} is increased monotonically from 0 T. The specific Dirichlet boundary conditions on $\mathbf{u}_\mathcal{T}$ and $\varphi_\mathcal{T}$ in this step for the **F**-**H** model reads

$$\mathbf{u}_\mathcal{T} = \mathbf{0}, \ \forall\, \mathbf{X} \in \partial\mathcal{V}_{\mathrm{MRE}}^{\mathrm{Left}}, \ \text{ and } \mathbf{u}_\mathcal{T} = \mathbf{0}, \quad \forall\, \mathbf{X} \in \partial\mathcal{V}_{\mathrm{Air}} \tag{140}$$

$$\varphi_\mathcal{T} = 0, \ \forall\, \mathbf{X} \in \partial\mathcal{V}_{\mathrm{Air}}^{\mathrm{Bottom}}, \ \text{ and } \varphi_\mathcal{T} = -\frac{b_{2,\mathcal{T}}^{\mathrm{actu}}}{\mu_0}L, \quad \forall\, \mathbf{X} \in \partial\mathcal{V}_{\mathrm{Air}}^{\mathrm{Top}}. \tag{141}$$

In turn, the boundary condition on α in this step for the **F**-**B** model computations reads

$$\alpha_\mathcal{T} = 0, \ \forall\, \mathbf{X} \in \partial\mathcal{V}_{\mathrm{Air}}^{\mathrm{Left}}, \ \text{ and } \alpha_\mathcal{T} = -b_{2,\mathcal{T}}^{\mathrm{actu}}L, \ \forall\, \mathbf{X} \in \partial\mathcal{V}_{\mathrm{Air}}^{\mathrm{Right}}. \tag{142}$$

In addition, we choose to work with a particle volume fraction of $c = 0.177$, which is identical to that of the fabricated h-MREs by Zhao et al. (2019). Moreover, we consider $G_{\mathrm{m}} = 0.187$ MPa, which leads to an effective shear modulus $G = 0.303$ MPa for the composite. In fact, the latter is experimentally measured by Zhao et al. (2019) for the h-MREs with $c = 0.177$.

In agreement to the experimental observations, the computations show the pre-magnetized h-MREs to deflect immediately under the applied b_2^{actu}. The end-tip deflections of the pre-magnetized cantilevers with an increasing b_2^{actu} is plotted in Fig. 22a for $r_{\mathrm{asp}} = 10$ and 17.5. Therein, we observe that the **F**-**H** and **F**-**B**-based numerical simulations yield identical responses, which also agree with the experimentally measured end-tip deflection values for the two aforementioned aspect ratios. Moreover, the experimentally captured deflected shape in Fig. 22b, which is of the cantilever beam having $r_{\mathrm{asp}} = 10$ under $b_2^{\mathrm{actu}} = 25$ mT agrees excellently with its numerically computed counterpart in Fig. 22c. The FE solutions are carried out via writing an user-defined element (UEL) and coupling it with the ABAQUS/Standard solver.

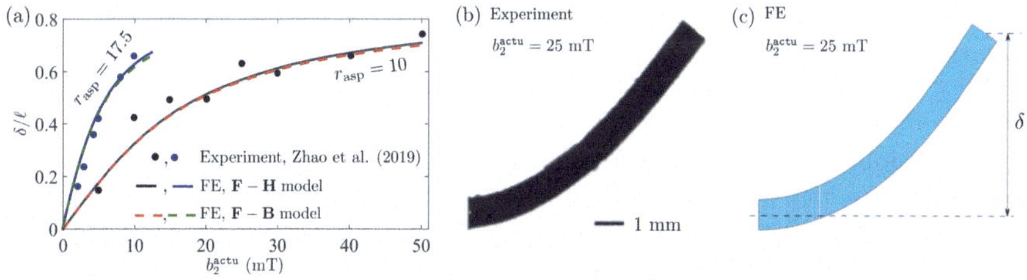

Fig. 22 **a** Comparison of the experimentally measured and the model (both **F**-**H** and **F**-**B**) predicted end-tip deflections of the pre-magnetized cantilever beams of $r_{\mathrm{asp}} = 10$ and 17.5 under the applied actuation field along $\mathbf{e}_2$. Comparison of the FE predicted deflected beam shape with the respective experimental measurements of Zhao et al. (2019) under $b_2^{\mathrm{actu}} = 25$ mT for **b** $r_{\mathrm{asp}} = 10$ and **c** 17.5. (Taken from Mukherjee and Danas (2022))

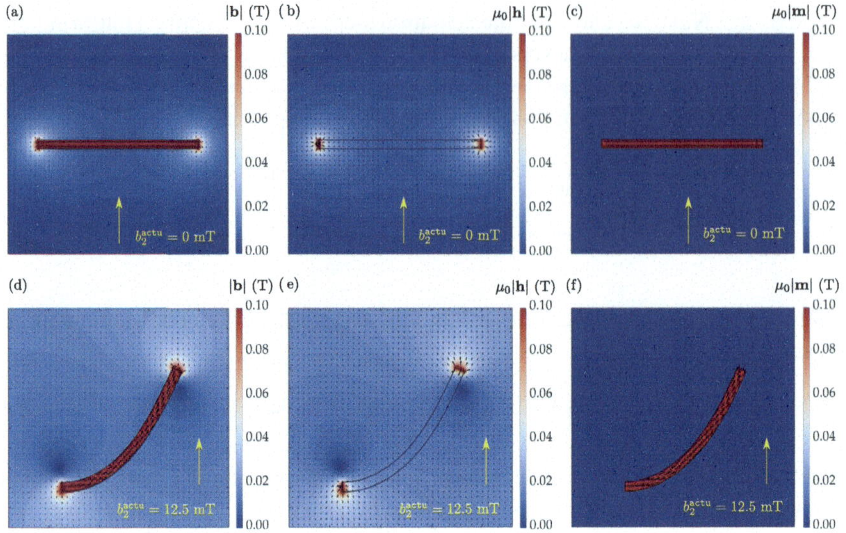

Fig. 23 Contours of the **a**, **d** $|\mathbf{b}|$, **b**, **e** $|\mathbf{h}|$ and **c**, **f** $|\mathbf{m}|$ fields in and around a pre-magnetized h-MRE cantilever of $r_{asp} = 17.5$, (**a–c**) before and (**d–f**) after the application of an actuation field $b_2^{actu} = 12.5$ mT along $\widehat{\mathbf{E}}_2$. The black colored arrows are used to indicate the direction of the respective vector fields. The length of the arrows are scaled according to the magnitude of the respective vectors. (Taken from Mukherjee and Danas (2022))

The contours of the magnetic **b**, **h** and **m** field magnitudes along with the arrows showing their directions in and around the pre-magnetized h-MRE cantilever of $r_{asp} = 17.5$ are shown in Fig. 23. Specifically, we show the contours under $b_2^{actu} = 0$ mT and $b_2^{actu} = 12.5$ mT in Fig. 23a–f, respectively. Notice from Fig. 23b that the **h** field in the pre-magnetized cantilever is considerably smaller than the **b** and **m** fields in it. Thus, one can approximate the remanent b-field,i.e., the b-field in the h-MRE after pre magnetization as shown in Fig. 23a, to be $\mathbf{b}^r \approx \mu_0\mathbf{m}$. This is, in fact, the key feature upon which the magnetic torque-based models (Kim et al., 2018; Zhao et al., 2019) for the pre-magnetized h-MREs are based. Such simple approximations, however, do not hold in general for the cases of non-uniform pre-magnetization or the hybrid h-/s-MRE beams. Specific examples of the hybrid hybrid h-/s-MRE beams and non-uniform pre-magnetization will be discussed later in this section.

The contours in and around the deflected h-MRE under $b_2^{actu} = 12.5$ mT in Fig. 23d–f show that the magnetic self fields (both, **b** and **h** but not **m**, which is 0 in the air) around it get perturbed by the external field application and the mechanical deformation of the beam. The remanent **b** and **m** fields in the h-MRE, however, only undergo rotation with a negligible change in their magnitudes. Clearly, the applied field $b_2^{actu} = 12.5$ mT, which results in such a rapid deflection of the cantilever, is too weak to alter the remanent magnetization direction. Thus, in spite of being a dissipative material in general, such very low field deflections of the pre-magnetized beams leads to a highly reversible structural response, hence, making them an ideal candidate for the remotely-actuated soft robots (Ren et al., 2019; Alapan et al., 2020; Lucarini et al., 2022a).

8.5 Non-uniformly Pre-magnetized, Functionally-Graded h-MRE Cantilever Beams

The increasing trend in the development of remotely-actuated locomotion of mili-robotic structures necessitates the employment of non-uniformly pre-magnetized h-MREs, exhibiting preferential deflections patterns depending on the actuation field directions (Ren et al., 2019; Alapan et al., 2020). Motivated from these recent applications, we employ the proposed incremental variational framework in the investigation of non-uniformly pre-magnetized h-MREs, specifically towards their pre-magnetization patterns and actuation performances.

In particular, we consider a slender h-MRE beam with $r_{\mathrm{asp}} = 20$, $G_{\mathrm{m}} = 0.187$ MPa and profile it according to the configurations shown in Fig. 24a, b before applying the pre-magnetization field along $\widehat{\mathbf{E}}_2$. Depending on this pre-magnetization profiling, the h-MREs are categorized into two, namely, $\mathcal{S}_1$ and $\mathcal{S}_2$, as indicated on Fig. 24a, b. Moreover, we consider two more types of h-MREs, namely, $\mathcal{T}_1$ and $\mathcal{T}_2$, depending on the spatial distribution of the particle volume fraction. In particular, we consider a constant $c = 0.177$ for type $\mathcal{T}_1$, while a linearly varying c along the reference coordinate X_1 so that $c = 0.054 + 0.492|X_1|/\ell$ for the type $\mathcal{T}_2$ (see Fig. 24c). Notice that the *cumulative volume* of the hard-magnetic particles are considered to be identical in $\mathcal{T}_1$ and $\mathcal{T}_2$, so that the areas under both the curves in Fig. 24c remain identical. Hence, we investigate the transverse actuation response of four distinct pre-magnetized h-MREs, namely, $\mathcal{S}_i\mathcal{T}_j$, where $i, j \equiv 1, 2$.

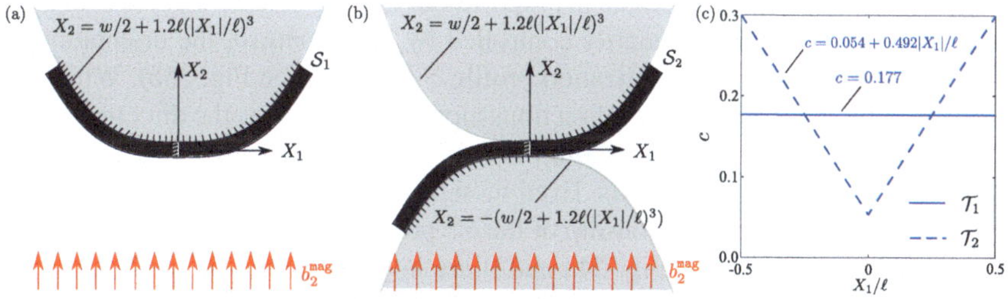

Fig. 24 **a, b** Pre-magnetization profiles along the magnetizing field b_2^{mag} direction for the h-MRE beams of $r_{\mathrm{asp}} = 20$. The profile $\mathcal{S}_1$ in (**a**) is considered to be the mirror image with respect to X_2 axis, whereas, $\mathcal{S}_2$ in (**b**) is considered to be the mirror image with respect to both, X_1 and X_2. **c** Volume fraction distribution profiles, namely, $\mathcal{T}_1$ and $\mathcal{T}_2$, along the length of the beam. (Taken from Mukherjee and Danas (2022))

Evidently, the initial profiling of the h-MRE beams and their release after the pre-magnetization necessitates a couple of additional steps of mechanical loading compared to the examples presented in Sects. 8.4. These steps read

- `Step-I`: First, the profiling of the undeformed to the pre-magnetization shapes are performed by applying a prescribed displacement $\mathbf{u}_\tau = \mathbf{u}_\tau^{\text{Top}}$ for all $\mathbf{X} \in \partial\mathcal{V}_{\text{MRE}}^{\text{Top}}$ for $\mathcal{S}_1$ and $\mathbf{u}_\tau = \mathbf{u}_\tau^{\text{Top}}$ for all $\mathbf{X} \in \partial\mathcal{V}_{\text{MRE}}^{\text{Top}}$, $X_1 > 0$ and $\mathbf{u}_\tau = \mathbf{u}_\tau^{\text{Bottom}}$ for all $\mathbf{X} \in \partial\mathcal{V}_{\text{MRE}}^{\text{Bottom}}$, $X_1 < 0$ for $\mathcal{S}_2$.[8] In addition, we set $\mathbf{u}_\tau = \mathbf{0}$ for all $\mathbf{X} \in \partial\mathcal{V}_{\text{Air}}$ and for all $\mathbf{X} \in \mathcal{V}_{\text{MRE}}$ if $X_1 = 0$, i.e., the displacements of the central vertical section of the beam are also blocked.
- `Step-II`: Next, the pre-magnetization is carried out along $\widehat{\mathbf{E}}_2$ in terms of applying a suitable Dirichlet boundary condition on φ_τ similar to Sect. 8.4. Moreover, the temporal profile and amplitude of $(b_2^{\mathtt{mag}})_\tau$ remains identical to that of $(b_1^{\mathtt{mag}})_\tau$ in Sect. 8.4.
- `Step-III`: This step gradually releases the constraints on $\mathbf{u}_\tau$ for all $\mathbf{X} \in \partial\mathcal{V}_{\text{MRE}}^{\text{Top}}$ and $\mathbf{X} \in \partial\mathcal{V}_{\text{MRE}}^{\text{Bottom}}$, while keeping $\mathbf{u}_\tau = \mathbf{0}$ at $\mathcal{V}_{\text{Air}}$ and the central vertical section of the beam. The beam comes back to its (almost) undeformed shape after this step.
- `Step-IV`: This is essentially the actuation step where the field $(b_2^{\mathtt{actu}})_\tau$ is applied along $\widehat{\mathbf{E}}_2$. Hence, the Dirichlet boundary condition on φ_τ is set identical to (141), while that on $\mathbf{u}_\tau$ remain the same as at the end of `Step-III`.

The first key outcome from the aforementioned magneto-mechanical loading exercise is the variation of the remanent magnetization $\mathbf{m}^0$ along the beam's centerline at the end of `Step-III`. Specifically, the variation of the magnitude of $\mathbf{m}^0$ and its angle with $\widehat{\mathbf{E}}_1$ for all four combinations of pre-magnetization and c profiles, namely, $\mathcal{S}_i\mathcal{T}_j$ with $i, j = 1, 2$ are shown in Fig. 25a, b, respectively. In agreement with the experimental observations (Ren et al., 2019; Alapan et al., 2020) , the magnitude of $\mathbf{m}^0$ remains the same in the beams $\mathcal{S}_1\mathcal{T}_1$ and $\mathcal{S}_2\mathcal{T}_1$, which have a spatially uniform c. The beams $\mathcal{S}_1\mathcal{T}_2$ and $\mathcal{S}_2\mathcal{T}_2$, on the other hand, exhibit a variation of $|\mathbf{m}^0|$ along the centerline. In fact this variation is proportional to the c variation in these beams. Thus, $|\mathbf{m}^0|$ in the beam is primarily controlled by c. In contrast, the orientation of $\mathbf{m}^0$ is dictated by its pre-magnetization profile $\mathcal{S}_1$ and $\mathcal{S}_2$ (see Fig. 25b). While the $\mathcal{S}_1$-type beams show opposite $\mathbf{m}^0$ directions along its two flanks, the direction of $\mathbf{m}^0$ in $\mathcal{S}_2$-type beams are identical in both the flanks, hence, showing a *bell curve* like variation in angle with the X_1 axis. Even though $|\mathbf{m}^0|$ in the beam is predictable in terms of c, the functional relationship of the θ^{m} profiles in Fig. 25b with the respective pre-deformed shapes in Fig. 24a and b are not straightforward and cannot be predicted beforehand prior solving the full field BVP.

---

[8] In practice, we employ the "DISP" subroutine of ABAQUS, which apply an user-defined displacement in terms of the *current coordinates*. We thus define the displacements $u_1^{\text{Top}} = -0.6\ell(x_1/\ell)^3$ and $u_2^{\text{Top}} = 1.2\ell(|x_1|/\ell)^3$ for $\mathcal{S}_1$ and additionally, $u_1^{\text{Bottom}} = -0.6\ell(x_1/\ell)^3$ and $u_2^{\text{Bottom}} = -1.2\ell(x_1/\ell)^3$ to achieve the deformation profile $\mathcal{S}_2$. These displacements are applied incrementally, held to the prescribed constant values and then released incrementally during the appropriate steps.

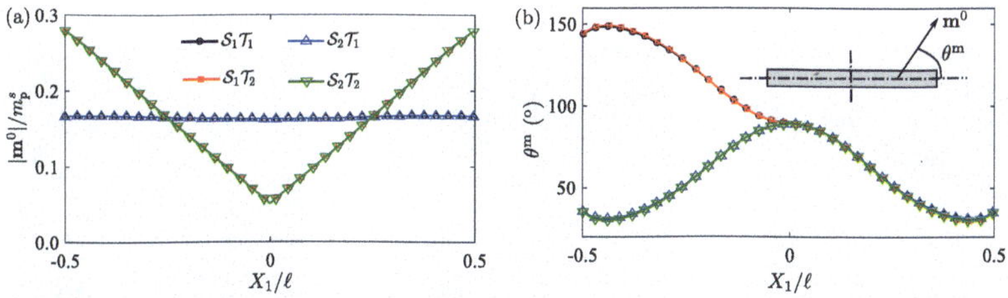

Fig. 25 Variation of the remanent magnetization $\mathbf{m}^0$ **a** magnitudes and **b** directions along the length of the pre-magnetized h-MRE beams of type $\mathcal{S}_i\mathcal{T}_i$ $(i = 1, 2)$. (Taken from Mukherjee and Danas (2022))

To obtain more insight on the complexity of the non-uniform remanent fields, we plot the spatial profiles of $\mathbf{b}^0$, $\mathbf{h}^0$ and $\mathbf{m}^0$ fields, both, in terms of magnitude and directions, in Fig. 26 for all four aforementioned types of beams. The first, and obvious feature observed is the higher magnitude of $\mathbf{b}^0$, $\mathbf{h}^0$ and $\mathbf{m}^0$ in the beams of type $\mathcal{T}_2$, which can directly be attributed to the higher c value in $\mathcal{T}_2$ near the beam flanks (cf. e.g., Fig. 26a–c, g–i). Moreover, comparing Fig. 26b, e with h, k we observe that by linearly increasing c along the flanks, the concentration of $\mathbf{h}^0$ field near the center of the beam can be eliminated. Of course, the spatial gradient of c in the $\mathcal{T}_2$-type beams results in a stiffer gradient in the $|\mathbf{b}|^0$ along the beams's centerline (cf. Fig. 26a, d, g, j).

In contrast, the directions of $\mathbf{b}^0$, $\mathbf{h}^0$ and $\mathbf{m}^0$ fields in the h-MRE along with the stray fields around the MRE domain depend strongly on their pre-magnetization profiles. Thus, we observe *qualitatively similar* stray $\mathbf{b}$ and $\mathbf{h}$ field distributions around the in all the $\mathcal{S}_1$ or $\mathcal{S}_2$-type beams, irrespective of the c distributions in them. Specifically, we observe from Fig. 26a, g that the beams with pre-magnetization profile $\mathcal{S}_1$ exhibit stronger self fields at the vicinity of their bottom boundary as compared to the top. Such preferential self-field distributions are typically achieved by constructing *Halbach chains* (Halbach, 1980; Hilton & McMurry, 2012; Mansson, 2014), which consists of an array of permanent magnets arranged in a particular fashion in order to concentrate the resulting magnetic self field at one side of the chain. A similar feature is observed here for the $\mathcal{S}_1$-type non-uniformly pre-magnetized hMRE in Fig. 26a, g. Thus, a properly pre-magnetized, *monolithic* h-MRE can mimic the properties of classical, essentially heterogeneous, Halbach chain structures.

In turn, such a concentration of the magnetic self fields are not observed in the beams having the pre-magnetization profile as $\mathcal{S}_2$. Rather, the contours of higher magnetic self fields render an inverted "S"-type shapes in all the $\mathcal{S}_2$-type beams. Thus, proper profiling of the beam before the pre-magnetization may help engineering different self field distributions in the vicinity of a h-MRE. The implications of such self field distributions on the actuation response of the $\mathcal{S}_1$ and $\mathcal{S}_2$-type beams will be discussed next.

Remark 8.1 The $\mathbf{b}$, $\mathbf{h}$ and $\mathbf{m}$ field magnitude contours and directions in Fig. 26 reveal that the local remanent $\mathbf{b}$ and $\mathbf{m}$ fields in the h-MREs *are not* related by

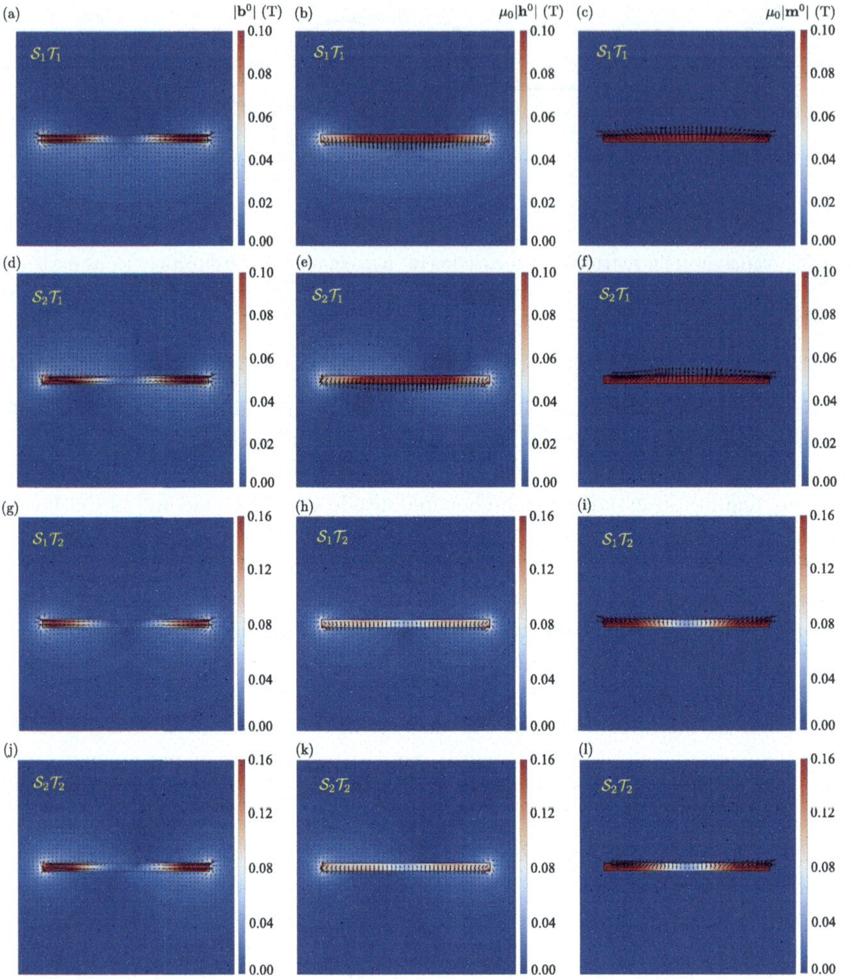

Fig. 26 Magnitude contours along with the directions of the remanent $\mathbf{b}^0$, $\mathbf{h}^0$ and $\mathbf{m}^0$ fields *after* the pre-magnetization step (Step-3) for the h-MRE types (depending on the pre-deformation profile and particle distribution) **a–c** $\mathcal{S}_1\mathcal{T}_1$, **d–f** $\mathcal{S}_2\mathcal{T}_1$, **g–i** $\mathcal{S}_1\mathcal{T}_2$ and **j–l** $\mathcal{S}_2\mathcal{T}_2$. (Taken from Mukherjee and Danas (2022))

the relation $\mathbf{b} = \mu_0\mathbf{m}$. Hence, unlike the uniformly pre-magnetized h-MREs, it may reveal inaccurate to assume, in general, that the magnetic torque at a point in the h-MRE is simply given by $\mathbf{b}^{\text{actu}} \times \mu_0\mathbf{m}$ during the actuation under remotely applied $\mathbf{b}^{\text{actu}}$ field. Thus, even though the magnetic toque-based, reduced-order models for slender h-MRE beams exhibit sufficiently accurate deflection profiles (Wang et al., 2020; Yan et al., 2021a, b), their employment to the non-uniformly pre-magnetized h-MRE structures must be carried out with caution and certainly use the local non-uniform pre-magnetization profile. Moreover, the pre-magnetization directions

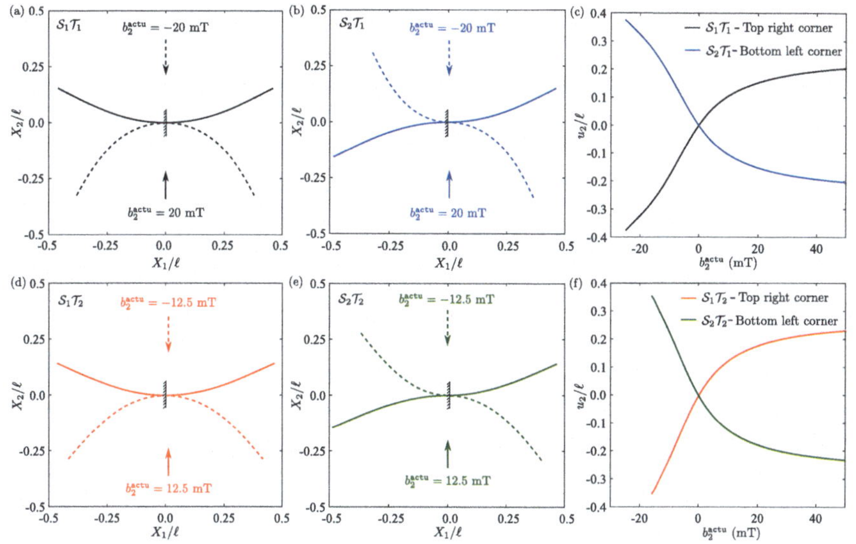

Fig. 27 Deflected centerlines of the h-MREs type **a** $\mathcal{S}_1\mathcal{T}_1$ and **b** $\mathcal{S}_2\mathcal{T}_1$ under the actuation fields $b_2^{\mathrm{actu}} = 20$ and -20 mT along $\widehat{\mathbf{E}}_2$. **c** Variation of the top-right corner delfection of $\mathcal{S}_1\mathcal{T}_1$ and bottom left corner delfection $\mathcal{S}_2\mathcal{T}_1$ under b_2^{actu}. Deflected centerlines of the h-MREs type **d** $\mathcal{S}_1\mathcal{T}_2$ and **e** $\mathcal{S}_2\mathcal{T}_2$ under the actuation fields $b_2^{\mathrm{actu}} = 12.5$ and -12.5 mT along $\widehat{\mathbf{E}}_2$. **f** Variation of the top-right corner delfection of $\mathcal{S}_1\mathcal{T}_2$ and bottom left corner delfection $\mathcal{S}_2\mathcal{T}_2$ under b_2^{actu}. (Taken from Mukherjee and Danas (2022))

along the beam length do not exhibit any straightforward correlation with its pre-deformation geometry. Hence, solving for the full-field BVP with a surrounding air becomes inevitable even for a reduced-order analysis in the later stage.

Finally, we show the transverse magnetic actuation performance of the four types of beams, namely, $\mathcal{S}_i\mathcal{T}_j$ with $i, j \equiv 1, 2$. First, we investigate the uniformly distributed c cases, i.e., the response of $\mathcal{T}_1$-type h-MREs in Fig. 27a–c. Specifically, Fig. 27a, b show the deflected beam centerline under an actuation field $b_2^{\mathrm{actu}} = 20$ mT along $\widehat{\mathbf{E}}_2$ and $-\widehat{\mathbf{E}}_2$ directions. Identical deflection of both the beam flanks are observed for the beam $\mathcal{S}_1\mathcal{T}_1$. However, the deflection is substantially higher ($\sim$2.5 times) when the fields are applied along $-\widehat{\mathbf{E}}_2$. This observation can directly be attributed to the pre-magnetization direction in both the beam flanks, which, eventually leads to a higher deflection when deflecting in the opposite direction of $\mathbf{m}^0$. This preferential deflection phenomena can be termed as the *magneto-mechanical Halbach effect*. In fact, this preferential deflection property is harnessed effectively in locomotion of soft jellyfish-like swimming robot (Ren et al., 2019; Alapan et al., 2020).

Even though equal in their magnitudes, the two flanks of the $\mathcal{S}_2$-type pre-magnetized beams always deflect in the opposite direction. For example, the deflected centerline of the h-MRE beam under $b_2^{\mathrm{actu}} = 20$ mT along $\widehat{\mathbf{E}}_2$ and $-\widehat{\mathbf{E}}_2$ are shown in Fig. 27b. Of course, here the deflection magnitude remains identical to the $\mathcal{S}_1$-type h-MRE, but overall, leading to a rocking-type motion, where the two flanks deflect simultaneously in the opposite directions. In particular, the variation of the vertical displacement of top-right and bottom-left corners of the $\mathcal{S}_1\mathcal{T}_1$ and $\mathcal{S}_2\mathcal{T}_1$-type h-MRE beams under $b_2^{\mathtt{actu}}$ are shown in Fig. 27c. This figure clearly shows that the deflection magnitude under the same $|b_2^{\mathtt{actu}}|$ becomes $\sim$2.5 times when the direction of its application is along $-\widehat{\mathbf{E}}_2$.

The $\mathcal{T}_2$-type beams exhibit a qualitatively similar deflection response under $b_2^{\mathtt{actu}}$ *vis-a-vis* the $\mathcal{T}_1$-type. The only and obvious difference between the former and the latter is that the $\mathcal{T}_2$-type beams deflect the same amount at a lower actuation field ($\sim$0.75 times). The deflected shapes of the $\mathcal{S}_1\mathcal{T}_2$ and $\mathcal{S}_2\mathcal{T}_2$-type beams are shown in Fig. 27d, e, respectively, both, under $b_2^{\mathtt{actu}} = 12.5$ and -12.5. Finally, the deflection variation of the top-right and bottom-left corners of, respectively, $\mathcal{S}_1\mathcal{T}_2$ and $\mathcal{S}_2\mathcal{T}_2$-type beams under $b_2^{\mathtt{actu}}$ are shown in Fig. 27f.

In closing, we remark that except lowering of the actuation field magnitude, the functionally-graded h-MREs with a linearly increasing c towards the beam flanks do not exhibit any substantial difference with the actuation performance of its uniform c counterpart. In turn, fabricating the functionally-graded h-MREs adds on to the difficulty level and cost. The pre-magnetization profiling, in contrast, can dramatically change the actuation performance of the h-MRE beams. In this regard, the proposed model serves as an efficient tool to analyze the effect of different pre-magnetization profiles and directions even before the manufacturing of an actual sample is carried out.

Acknowledgements The author would like to acknowledge support from the European Research Council (ERC) under the European Union's Horizon 2020 and Horizon Europe research and innovation program (grant agreement No. 636903—MAGNETO and No. 101081821—MagnetoSense). The computational part of this work was also supported by the ANR, France under contract number ANR-10-EQPX-37. The Abaqus user defined element (UEL) developed for the numerical simulations using the proposed **F**-**H** model along with a representative example are available at https://www.doi.org/10.5281/zenodo.4588578. The **F**-**B** model UEL will be made available upon request. The author would like to thank his wonderful students D. Mukherjee, M. Rambausek, and colleagues L. Bodelot, V. Lefèvre, O. Lopez-Pamies and N. Triantafyllidis for their important contributions, feedback and collaboration over the last 15 years that allowed to develop this rather wide class of MRE models.

References

ABAQUS. (2017). *CAE/2017*. RI, USA: Simulia.

Abramowitz, M., & Stegun, I. A. (1972). *Handbook of mathematical functions*. New York: Dover.

Alapan, Y., Karacakol, A. C., Guzelhan, S. N., Isik, I., & Sitti, M. (2020). Reprogrammable shape morphing of magnetic soft machines. *Science Advances, 6*(38), eabc6414. https://doi.org/10.1126/sciadv.abc6414.

Anoukou, K., Brenner, R., Hong, F., Pellerin, M., & Danas, K. (2018). Random distribution of polydisperse ellipsoidal inclusions and homogenization estimates for porous elastic materials.

Computers and Structures, 210, 87 – 101. ISSN 0045-7949. https://doi.org/10.1016/j.compstruc.2018.08.006. http://www.sciencedirect.com/science/article/pii/S0045794918304565.

Bastola, A. K., & Hossain, M. (2020). A review on magneto-mechanical characterizations of magnetorheological elastomers. *Composites Part B: Engineering, 200*, 108348. ISSN 1359-8368. https://doi.org/10.1016/j.compositesb.2020.108348. https://www.sciencedirect.com/science/article/pii/S1359836820333977.

Bennett, K. C., Regueiro, R. A., & Borja, R. I. (2016). Finite strain elastoplasticity considering the eshelby stress for materials undergoing plastic volume change. *International Journal of Plasticity,77*, 214–245. https://doi.org/10.1016/j.ijplas.2015.10.007.

Biro, O., & Preis, K. (1989). On the use of the magnetic vector potential in the finite-element analysis of three-dimensional eddy currents. *IEEE Transactions on Magnetics,25*(4), 3145–3159. https://doi.org/10.1109/20.34388.

Bodelot, L., Voropaieff, J.-P., & Pössinger, T. (2017). Experimental investigation of the coupled magneto-mechanical response in magnetorheological elastomers. *Experimental Mechanics, 58*(2), 207–221. https://doi.org/10.1007/s11340-017-0334-7.

Brown, W. F. (1966). *Magnetoelastic Interactions, 9*. Springer.

Bustamante, R., Dorfmann, A., & Ogden, R. W. (2008). On variational formulations in nonlinear magnetoelastostatics. *Mathematics and Mechanics of Solids,13*(8), 725–745. https://doi.org/10.1177/1081286507079832.

Chen, Y.-C., & Wheeler, L. (1993). Derivatives of the stretch and rotation tensors. *Journal of Elasticity,32*(3), 175–182. https://doi.org/10.1007/bf00131659.

Danas, K., & Triantafyllidis, N. (2014). Instability of a magnetoelastic layer resting on a non-magnetic substrate. *Journal of the Mechanics and Physics of Solids,69*, 67–83. https://doi.org/10.1016/j.jmps.2014.04.003.

Danas, K., Kankanala, S. V., & Triantafyllidis, N. (2012). Experiments and modeling of iron-particle-filled magnetorheological elastomers. *Journal of the Mechanics and Physics of Solids, 60*(1), 120–138. ISSN 0022-5096. https://doi.org/10.1016/j.jmps.2011.09.006. http://www.sciencedirect.com/science/article/pii/S0022509611001736.

Danas, K. (2017). Effective response of classical, auxetic and chiral magnetoelastic materials by use of a new variational principle. *Journal of the Mechanics and Physics of Solids,105*, 25–53. https://doi.org/10.1016/j.jmps.2017.04.016.

Dashner, P. A. (1993). An objective kinematical formalism for the modeling of elastic-plastic materials subject to large deformation. *International Journal of Solids and Structures,30*(19), 2661–2672. https://doi.org/10.1016/0020-7683(93)90104-f.

Deng, X., Liu, Z., Hongya, Y., Xiao, Z., & Zhang, G. (2015). Isotropic and anisotropic nanocrystalline NdFeB bulk magnets prepared by binder-free high-velocity compaction technique. *Journal of Magnetism and Magnetic Materials,390*, 26–30. https://doi.org/10.1016/j.jmmm.2015.04.075.

DeSimone, A., & James, R. D. (2002). A constrained theory of magnetoelasticity. *Journal of the Mechanics and Physics of Solids,50*(2), 283–320. https://doi.org/10.1016/s0022-5096(01)00050-3.

Diguet, G. (2010). *Huge Magnetostriction of Magneto-Rheological Composite*. Ph.D. thesis, Université de Grenoble.

Diguet, G., Beaugnon, E., & Cavaillé, J. Y. (2010). Shape effect in the magnetostriction of ferromagnetic composite. *Journal of Magnetism and Magnetic Materials, 322*(21), 3337–3341. ISSN 0304-8853. https://doi.org/10.1016/j.jmmm.2010.06.020. http://www.sciencedirect.com/science/article/B6TJJ-5093N8V-2/2/d5878c0abd20adc88cdb94281950f912.

Diguet, G., Sebald, G., Nakano, M., Lallart, M., & Cavaillé, J.-Y. (2021). Magnetic behavior of magneto-rheological foam under uniaxial compression strain. *Smart Materials and Structures,31*(2), 025018. https://doi.org/10.1088/1361-665x/ac3fc8. https://doi.org/10.1088%2F1361-665x%2Fac3fc8.

Dorfmann, A., & Ogden, R. W. (2004). Nonlinear magnetoelastic deformations of elastomers. *Acta Mechanica,167*(1–2), 13–28. https://doi.org/10.1007/s00707-003-0061-2.

Dorn, C., Bodelot, L., & Danas, K. (2021). Experiments and numerical implementation of a boundary value problem involving a magnetorheological elastomer layer subjected to a nonuniform magnetic field. *Journal of Applied Mechanics, 88*(7). https://doi.org/10.1115/1.4050534.

Eringen, A. C. (1967). *Mechanics of continua*. New York: Wiley.

Eringen, A. C., & Maugin, G. A. (1990). *Electrodynamics of continua I: Foundations and solid media*. New York: Springer.

Galipeau, E., & Castañeda, P. P. (2013). A finite-strain constitutive model for magnetorheological elastomers: Magnetic torques and fiber rotations. *Journal of the Mechanics and Physics of Solids,61*(4), 1065–1090. https://doi.org/10.1016/j.jmps.2012.11.007.

Garcia-Gonzalez, D., Ter-Yesayants, T., Moreno-Mateos, M. A., & Lopez-Donaire, M. L. (2023). Hard-magnetic phenomena enable autonomous self-healing elastomers. *Composites Part B: Engineering, 248*, 110357, 2023. ISSN 1359-8368. https://doi.org/10.1016/j.compositesb.2022.110357. https://www.sciencedirect.com/science/article/pii/S1359836822007302.

Gebhart, P., & Wallmersperger, T. (2022a). A constitutive macroscale model for compressible magneto-active polymers based on computational homogenization data: Part I—magnetic linear regime. *International Journal of Solids and Structures, 236-237*, 111294. ISSN 0020-7683. https://doi.org/10.1016/j.ijsolstr.2021.111294. https://www.sciencedirect.com/science/article/pii/S0020768321003747.

Gebhart, P., & Wallmersperger, T. (2022b). A constitutive macroscale model for compressible magneto-active polymers based on computational homogenization data: Part II—Magnetic nonlinear regime. *International Journal of Solids and Structures, 258*, 111984. ISSN 0020-7683. https://doi.org/10.1016/j.ijsolstr.2022.111984. https://www.sciencedirect.com/science/article/pii/S0020768322004371.

Ginder, J. M., Nichols, M. E., Elie, L. D., & Tardiff, J. L. (1999). Magnetorheological elastomers: Properties and applications. In *Smart Structures and Materials 1999: Smart Materials Technologies* (Vol. 3675, pp. 131–139). International Society for Optics and Photonics. https://doi.org/10.1117/12.352787.

Gurtin, M. E. (1982). *An introduction to continuum mechanics*. Academic Press.

Halbach, K. (1980). Design of permanent multipole magnets with oriented rare earth cobalt material. *Nuclear Instruments and Methods,169*(1), 1–10. https://doi.org/10.1016/0029-554x(80)90094-4.

Halphen, B., & Nguyen, Q. S. (1975). Sur les matériaux standard généralisés. *Journal de Mécanique, 14*, 39–63.

Hankin, R. K. S. (2015). Numerical evaluation of the gauss hypergeometric function with the hypergeo package. *The R Journal, 7*, 81–88.

Hilton, J. E., & McMurry, S. M. (2012). An adjustable linear halbach array. *Journal of Magnetism and Magnetic Materials, 324*(13):2051–2056. ISSN 0304-8853. https://doi.org/10.1016/j.jmmm.2012.02.014. https://www.sciencedirect.com/science/article/pii/S030488531200100X.

Idiart, M. I., Danas, K., & Castañeda, P. P. (2006a). Second-order theory for nonlinear composites and application to isotropic constituents. *Comptes Rendus Mécanique, 334*(10), 575–581. ISSN 1631-0721. https://doi.org/10.1016/j.crme.2006.06.006. https://www.sciencedirect.com/science/article/pii/S1631072106001070.

Idiart, M. I., Moulinec, H., Castañeda, P. P., & Suquet, P. (2006b). Macroscopic behavior and field fluctuations in viscoplastic composites: second-order estimates versus full-field simulations. *Journal of the Mechanics and Physics of Solids, 54*(5):1029–1063. https://doi.org/10.1016/j.jmps.2005.11.004. http://www.sciencedirect.com/science/article/pii/S0022509605002188.

James, R. D., & Kinderlehrer, D. (1993). Theory of magnetostriction with applications to tbxdy1-xfe2. *Phil. Mag. B, 68*, 237–274. ISSN 0141-8637.

Jolly, M. R., Carlson, J. D., Muñoz, B. C., & Bullions, T. A. (1996). The magnetoviscoelastic response of elastomer composites consisting of ferrous particles embedded in a polymer matrix. *Journal of Intelligent Material Systems and Structures, 7*(6), 613–622. https://doi.org/10.1177/1045389x9600700601.

Kalina, K. A., Brummund, J., Metsch, P., Kästner, M., Borin, D. Y., Linke, J. M., & Odenbach, S. (2017). Modeling of magnetic hystereses in soft MRES filled with NDFEB particles. *Smart Materials and Structures, 26*(10), 105019. https://doi.org/10.1088/1361-665x/aa7f81.

Kankanala, S. V., & Triantafyllidis, N. (2004). On finitely strained magnetorheological elastomers. *Journal of the Mechanics and Physics of Solids,52*(12), 2869–2908. https://doi.org/10.1016/j.jmps.2004.04.007.

Keip, M.-A., & Sridhar, A. (2018). A variationally consistent phase-field approach for micromagnetic domain evolution at finite deformations. *Journal of the Mechanics and Physics of Solids*. https://doi.org/10.1016/j.jmps.2018.11.012.

Keip, M.-A., & Rambausek, M. (2016). A multiscale approach to the computational characterization of magnetorheological elastomers. *International Journal for Numerical Methods in Engineering,107*(4), 338–360. https://doi.org/10.1002/nme.5178.

Kim, Y., Yuk, H., Zhao, R., Chester, S. A., & Zhao, X. (2018). Printing ferromagnetic domains for untethered fast-transforming soft materials. *Nature,558*(7709), 274–279. https://doi.org/10.1038/s41586-018-0185-0.

Kumar, A., & Lopez-Pamies, O. (2016). On the two-potential constitutive modeling of rubber viscoelastic materials. *Comptes Rendus Mécanique,344*(2), 102–112. https://doi.org/10.1016/j.crme.2015.11.004.

Kumar, S., Danas, K., & Kochmann, D. M. (2019). Enhanced local maximum-entropy approximation for stable meshfree simulations. *Computer Methods in Applied Mechanics and Engineering,344*, 858–886. https://doi.org/10.1016/j.cma.2018.10.030.

Lefèvre, V., Danas, K., & Lopez-Pamies, O. (2017). A general result for the magnetoelastic response of isotropic suspensions of iron and ferrofluid particles in rubber, with applications to spherical and cylindrical specimens. *Journal of the Mechanics and Physics of Solids,107*, 343–364. https://doi.org/10.1016/j.jmps.2017.06.017.

Lefèvre, V., Danas, K., & Lopez-Pamies, O. (2019). Two families of explicit models constructed from a homogenization solution for the magnetoelastic response of MRES containing iron and ferrofluid particles. *International Journal of Non-Linear Mechanics*. ISSN 0020-7462. https://doi.org/10.1016/j.ijnonlinmec.2019.103362. http://www.sciencedirect.com/science/article/pii/S0020746219306237.

Lopez-Pamies, O., Goudarzi, T., & Danas, K. (2013). The nonlinear elastic response of suspensions of rigid inclusions in rubber: II—a simple explicit approximation for finite-concentration suspensions. *Journal of the Mechanics and Physics of Solids,61*(1), 19–37. https://doi.org/10.1016/j.jmps.2012.08.013.

Lucarini, S., Moreno-Mateos, M. A., Danas, K., & Garcia-Gonzalez, D. (2022a). Insights into the viscohyperelastic response of soft magnetorheological elastomers: Competition of macrostructural versus microstructural players. *International Journal of Solids and Structures, 256*, 111981. ISSN 0020-7683. https://doi.org/10.1016/j.ijsolstr.2022.111981. https://www.sciencedirect.com/science/article/pii/S0020768322004346.

Lucarini, S., Hossain, M., & Garcia-Gonzalez, D. (2022). Recent advances in hard-magnetic soft composites: Synthesis, characterisation, computational modelling, and applications. *Composite Structures,279*, 114800.

Mansson, Daniel. (2014). On the suitability of using halbach arrays as potential energy storage media. *Progress In Electromagnetics Research B,58*, 151–166. https://doi.org/10.2528/pierb14010704.

McMeeking, R. M., & Landis, C. M. (2005). Electrostatic forces and stored energy for deformable dielectric materials. *Journal of Applied Mechanics,72*(4), 581–590. https://doi.org/10.1115/1.1940661.

McMeeking, R. M., Landis, C. M., & Jimenez, S. M. A. (2007). A principle of virtual work for combined electrostatic and mechanical loading of materials. *International Journal of Non-Linear Mechanics, 42*(6), 831–838. https://doi.org/10.1016/j.ijnonlinmec.2007.03.008.

Michel, J.-C., Moulinec, H., & Suquet, P. (1999). Effective properties of composite material with periodic microstructure: A computational approach. *Computer Methods in Applied Mechanics and Engineering, 172*, 109–143.

Miehe, C., Rosato, D., & Kiefer, B. (2011). Variational principles in dissipative electro-magneto-mechanics: A framework for the macro-modeling of functional materials. *International Journal for Numerical Methods in Engineering,86*(10), 1225–1276. https://doi.org/10.1002/nme.3127.

Moreno-Mateos, M. A., Gonzalez-Rico, J., Nunez-Sardinha, E., Gomez-Cruz, C., Lopez-Donaire, M. L., Lucarini, S., Arias, A., Muñoz-Barrutia, A., Velasco, D., & Garcia-Gonzalez, D. (2022). Magneto-mechanical system to reproduce and quantify complex strain patterns in biological materials. *Applied Materials Today, 27*, 101437. ISSN 2352-9407. https://doi.org/10.1016/j.apmt.2022.101437. https://www.sciencedirect.com/science/article/pii/S2352940722000762.

Mukherjee, D., & Danas, K. (2019). An evolving switching surface model for ferromagnetic hysteresis. *Journal of Applied Physics,125*(3), 033902. https://doi.org/10.1063/1.5051483.

Mukherjee, D. (2020). Theoretical and numerical modeling of magnetorheological elastomers comprising magnetically soft and hard particles. *Ph.D. thesis*.

Mukherjee, D., & Danas, K. (2022). A unified dual modeling framework for soft and hard magnetorheological elastomers. *International Journal of Solids and Structures, 257*, 111513. ISSN 0020-7683. https://doi.org/10.1016/j.ijsolstr.2022.111513. https://www.sciencedirect.com/science/article/pii/S0020768322000725. Special Issue in the honour Dr Stelios Kyriakides.

Mukherjee, D., Bodelot, L., & Danas, K. (2020). Microstructurally-guided explicit continuum models for isotropic magnetorheological elastomers with iron particles. *International Journal of Non-Linear Mechanics*, 103380. https://doi.org/10.1016/j.ijnonlinmec.2019.103380.

Mukherjee, D., Rambausek, M., & Danas, K. (2021). An explicit dissipative model for isotropic hard magnetorheological elastomers. *Journal of the Mechanics and Physics of Solids,151*, 104361. https://doi.org/10.1016/j.jmps.2021.104361.

Ogden, R. W. (1997). *Non-linear elastic deformations*. Courier Corporation.

Onsager, L. (1931a). Reciprocal relations in irreversible processes. I. *Physical Review, 37*(4), 405–426, February 1931a. https://doi.org/10.1103/physrev.37.405.

Onsager, L. (1931). Reciprocal relations in irreversible processes. II. *Physical Review,38*(12), 2265–2279. https://doi.org/10.1103/physrev.38.2265.

Papadioti, I., Danas, K., & Aravas, N. (2016). A methodology for the estimation of the effective yield function of isotropic composites. *International Journal of Solids and Structures, 87*, 120–138. ISSN 0020-7683. https://doi.org/10.1016/j.ijsolstr.2016.02.022. https://www.sciencedirect.com/science/article/pii/S0020768316000846.

Park, S., Mondal, K., Treadway, R. M., Kumar, V., Ma, S., Holbery, J. D., & Dickey, M. D. (2018). Silicones for stretchable and durable soft devices: Beyond sylgard-184. *ACS Applied Materials & Interfaces,10*(13), 11261–11268. https://doi.org/10.1021/acsami.7b18394.

Perger, W. F., Bhalla, A., & Nardin, M. (1993). A numerical evaluator for the generalized hypergeometric series. *Computer physics communications, 77*(2), 249–254.

Ponte Castañeda, P., & Galipeau, E. (2011). Homogenization-based constitutive models for magnetorheological elastomers at finite strain. *Journal of the Mechanics and Physics of Solids, 59*(2), 194–215. https://doi.org/10.1016/j.jmps.2010.11.004.

Psarra, E., Bodelot, L., & Danas, K. (2019). Wrinkling to crinkling transitions and curvature localization in a magnetoelastic film bonded to a non-magnetic substrate. *Journal of the Mechanics and Physics of Solids,133*, 103734. https://doi.org/10.1016/j.jmps.2019.103734.

Psarra, E., Bodelot, L., & Danas, K. (2017). Two-field surface pattern control via marginally stable magnetorheological elastomers. *Soft Matter,13*(37), 6576–6584. https://doi.org/10.1039/c7sm00996h.

Rambausek, M., & Keip, M.-A. (2018). Analytical estimation of non-local deformation-mediated magneto-electric coupling in soft composites. *Proceedings of the Royal Society A, 474*(2216), 20170803. ISSN 1364-5021, 1471-2946. https://doi.org/10.1098/rspa.2017.0803. http://rspa.royalsocietypublishing.org/content/474/2216/20170803.

Rambausek, M., Mukherjee, D., & Danas, K. (2022). A computational framework for magnetically hard and soft viscoelastic magnetorheological elastomers. *Computer Methods in Applied Mechanics and Engineering, 391*, 114500. ISSN 0045-7825. https://doi.org/10.1016/j.cma.2021.114500. https://www.sciencedirect.com/science/article/pii/S0045782521007064.

Rambausek, M., & Schöberl, J. (2023). Curing spurious magneto-mechanical coupling in soft non-magnetic materials. *International Journal for Numerical Methods in Engineering, 124*(10), 2261–2291. https://doi.org/10.1002/nme.7210. https://onlinelibrary.wiley.com/doi/abs/10.1002/nme.7210

Ren, Z., Hu, W., Dong, X., & Sitti, M. (2019). Multi-functional soft-bodied jellyfish-like swimming. *Nature Communications, 10*(1). https://doi.org/10.1038/s41467-019-10549-7.

Rosato, D., & Miehe, C. (2014). Dissipative ferroelectricity at finite strains. Variational principles, constitutive assumptions and algorithms. *International Journal of Engineering Science, 74*, 162–189. https://doi.org/10.1016/j.ijengsci.2013.08.007.

Sharma, B. L., & Saxena, P. (2020). Variational principles of nonlinear magnetoelastostatics and their correspondences. *Mathematics and Mechanics of Solids, 26*(10), 1424–1454. https://doi.org/10.1177/1081286520975808. https://doi.org/10.1177%2F1081286520975808

Stark, S., Semenov, A. S., & Balke, H. (2015). On the boundary conditions for the vector potential formulation in electrostatics. *International Journal for Numerical Methods in Engineering, 102*(11), 1704–1732. https://doi.org/10.1002/nme.4859.

Wang, L., Kim, Y., Guo, C. F., & Zhao, X. (2020). Hard-magnetic elastica. *Journal of the Mechanics and Physics of Solids, 142*, 104045. https://doi.org/10.1016/j.jmps.2020.104045.

Wang, Z., Xiang, C., Yao, X., Le Floch, P., Mendez, J., & Suo, Z. (2019). Stretchable materials of high toughness and low hysteresis. *Proceedings of the National Academy of Sciences, 116*(13), 5967–5972. https://doi.org/10.1073/pnas.1821420116.

Yan, D., Abbasi, A., & Reis, P. M. (2021a). A comprehensive framework for hard-magnetic beams: Reduced-order theory, 3d simulations, and experiments. *International Journal of Solids and Structures* (pp. 111319). ISSN 0020-7683. https://doi.org/10.1016/j.ijsolstr.2021.111319. https://www.sciencedirect.com/science/article/pii/S0020768321003978.

Yan, D., Pezzulla, M., Cruveiller, L., Abbasi, A., & Reis, P. M. (2021b). Magneto-active elastic shells with tunable buckling strength (Vol. 12(1)). https://doi.org/10.1038/s41467-021-22776-y. https://doi.org/10.1038%2Fs41467-021-22776-y.

Yan, D., Aymon, B. F. G., & Reis, P. M. (2023). A reduced-order, rotation-based model for thin hard-magnetic plates. *Journal of the Mechanics and Physics of Solids, 170*, 105095. ISSN 0022-5096. https://doi.org/10.1016/j.jmps.2022.105095. https://www.sciencedirect.com/science/article/pii/S0022509622002721.

Zhao, R., Kim, Y., Chester, S. A., Sharma, P., & Zhao, X. (2019). Mechanics of hard-magnetic soft materials. *Journal of the Mechanics and Physics of Solids, 124*, 244–263. https://doi.org/10.1016/j.jmps.2018.10.008.

Elastic Localizations

Yibin Fu

Abstract The past few decades have witnessed a surge of interest in pattern formations in soft materials under various fields. This has partly been driven by the recognition that buckling-induced patterns at micrometer and submicrometer scales may serve many useful purposes. Such patterns are usually either periodic or localized. Formation of periodic patterns is universally recognized as a bifurcation problem, and theories concerning periodic patterns have been well-developed and can be found in many textbooks and research monographs. In contrast, the initiation and evolution of localized patterns are rarely studied as a bifurcation problem, and when they are the discussion is often incomplete. In this chapter, we discuss three representative elastic localization problems: localized bulging of an inflated hyperelastic tube, localized necking of a solid cylinder induced by surface tension, and axisymmetric necking of a circular plate under all-round tension. All these problems are characterized by the fact that a linear bifurcation analysis would predict that the critical wavenumber is zero if the dimension in the direction of periodic variation is infinite. It is shown how the entire localization process, including initiation, growth and propagation, can be described analytically or semi-analytically, and how the process depends on how loading is carried out. It is also shown how a one-dimensional reduced model can be derived for the inflation problem and used to describe the entire localization process fairly accurately. It is hoped that the methodology explained here can be applied to study similar problems that also involve other effects such as electric and magnetic fields, chemical reactions, material deterioration, and residual stresses.

Y. Fu (✉)
School of Computer Science and Mathematics, Keele University, Keele, UK
e-mail: y.fu@keele.ac.uk

K. Danas and O. Lopez-Pamies (eds.), *Electro- and Magneto-Mechanics of Soft Solids*, CISM International Centre for Mechanical Sciences 610,
https://doi.org/10.1007/978-3-031-48351-6_4

1 An Example of Bifurcation at Zero Wavenumber

We first consider the simple boundary value problem (BVP)

$$u_{xx} + u_{yy} + Pu - u^2 = 0, \quad |y| < 1/2, \ |x| < \infty,$$

$$u(x, \pm 1/2) = 0, \tag{1}$$

where u is a function of x and y, P is a bifurcation parameter, and subscripts x and y signify partial differentiations. The trivial solution $u = 0$ is a solution for all values of P. Our interest is in finding the minimum value of P at which a non-trivial solution bifurcates from the trivial solution. The bifurcation behavior of this BVP has previously been studied by Kirchgässner (1982).

The linearized form of (1) admits a normal mode solution of the form

$$u = H(y)\mathrm{e}^{\mathrm{i}kx} + \text{c.c.}, \tag{2}$$

where c.c. denotes the complex conjugate of the preceding term, k is the wavenumber and the function $H(y)$ is determined by the eigenvalue problem

$$H''(y) + (P - k^2)H(y) = 0, \quad H(\pm 1/2) = 0. \tag{3}$$

It is straightforward to show that there exist two sets of non-trivial solutions, namely

$$H(y) = \cos(2n-1)\pi y, \quad \text{when} \ \ P = k^2 + (2n-1)^2\pi^2, \tag{4}$$

and

$$H(y) = \sin 2n\pi y, \quad \text{when} \ \ P = k^2 + 4n^2\pi^2, \tag{5}$$

where $n = 1, 2, \ldots$. Plotting $(4)_2$ and $(5)_2$ against k for $n = 1, 2, \ldots$ reveals that the lowest branch is given by

$$P = k^2 + \pi^2, \tag{6}$$

and corresponds to the *symmetric* mode $H(y) = \cos \pi y$. The minimum of P is π^2 and is attained at $k = 0$. Thus, as P is increased from 0, we expect there will be a non-trivial solution bifurcating from the trivial solution $u = 0$ at $P = \pi^2$.

To determine the near-critical behaviour of the bifurcated solution, we let

$$P = \pi^2 + \epsilon P_1, \tag{7}$$

where P_1 is an $O(1)$ constant and ϵ is a positive small parameter. Relation (6) implies that near-critical solutions have $k = O(\epsilon^{1/2})$. The linear solution then indicates that dependence of the solution on x should be through the product $\sqrt{\epsilon}x$. Thus, it is appropriate to define a far-distance variable X by

$$X = \sqrt{\epsilon}x. \tag{8}$$

Then, the u_{xx} in (1) becomes ϵu_{XX}. We would like this term to appear in the amplitude equation at second order (to satisfy a solvability condition). This can be achieved by requiring $\epsilon u_{XX} = O(u^2)$, from which we deduce that $u = O(\epsilon)$. Thus, we look for a near-critical perturbation solution of the form

$$u(x, y) = \epsilon u^{(1)}(X, y) + \epsilon^2 u^{(2)}(X, y) + \cdots . \tag{9}$$

On substituting (9) into (1) and equating the coefficients of ϵ and ϵ^2, we obtain

$$\mathcal{L}[u^{(1)}] = 0, \quad u^{(1)}(X, \pm 1/2) = 0, \tag{10}$$

and

$$\mathcal{L}[u^{(2)}] = -u^{(1)}_{XX} - P_1 u^{(1)} + (u^{(1)})^2, \quad u^{(2)}(X, \pm 1/2) = 0, \tag{11}$$

where the differential operator $\mathcal{L}$ is defined by $\mathcal{L}[u] \equiv u_{yy} + \pi^2 u$.

The leading-order problem (10) can easily be solved to yield

$$u^{(1)} = A(X) \cos \pi y, \tag{12}$$

where the amplitude function $A(X)$ is to be determined. On substituting (12) into $(11)_1$, we obtain

$$\mathcal{L}[u^{(2)}] = -(A_{XX} + P_1 A) \cos \pi y + A^2 \cos^2 \pi y. \tag{13}$$

The general solution of this inhomogeneous equation is given by

$$u^{(2)} = C_1(X) \cos \pi y + C_2(X) \sin \pi y + I(X, y), \tag{14}$$

where C_1, C_2 are arbitrary functions and the particular integral I is given by

$$I(X, y) = -\frac{1}{2\pi}(A_{XX} + P_1 A) y \sin \pi y + \frac{1}{2\pi^2} A^2 (1 - \frac{1}{3} \cos 2\pi y). \tag{15}$$

On substituting (14) into $(11)_2$, we obtain $C_2(X) + I(X, 1/2) = 0$ and $-C_2(X) + I(X, -1/2) = 0$. It then follows that $I(X, 1/2) + I(X, -1/2) = 0$, from which we obtain the amplitude equation

$$A_{XX} + P_1 A - \frac{8}{3\pi} A^2 = 0. \tag{16}$$

This equation admits a localized solution given by

$$A(X) = \frac{9\pi P_1}{16} \text{sech}^2 \left(\frac{1}{2} \sqrt{-P_1} X \right). \tag{17}$$

This solution exists only when $P_1 < 0$. Thus, at $P = P_{cr} = \pi^2$, a localized solution bifurcates sub-critically from the trivial solution $u = 0$.

Comment on solvability: Solvability conditions play an important role in post-bifurcation analysis, and can be understood using the matrix equation $A\boldsymbol{x} = \boldsymbol{b}$ as an example. This equation can be solved if A is invertible and the unique solution is $\boldsymbol{x} = A^{-1}\boldsymbol{b}$. However, if A is non-invertible, then this equation can be solved only if the right hand side $\boldsymbol{b}$ satisfies a solvability condition. To show this, we first note that in this case $A\boldsymbol{x} = \boldsymbol{0}$ must have a non-trivial solution, similar to (10) having a non-trivial solution. This non-trivial solution is a right eigenvector of A associated with the eigenvalue 0. There must also exist an associated left eigenvector, $\boldsymbol{y}$ say, such that $A^T\boldsymbol{y} = \boldsymbol{0}$. It then follows that $\boldsymbol{y} \cdot \boldsymbol{b} = \boldsymbol{y} \cdot A\boldsymbol{x} = \boldsymbol{x} \cdot A^T\boldsymbol{y} = \boldsymbol{0}$. The condition $\boldsymbol{y} \cdot \boldsymbol{b} = \boldsymbol{0}$ is the solvability condition for $A\boldsymbol{x} = \boldsymbol{b}$. We also note that if A is symmetric, then the right and left eigenvectors are the same. Extending this to differential operators such as the $\mathcal{L}$ above (which is self-adjoint), we may state that the BVP (11) can be solved only if a solvability condition is satisfied. This condition was found to be $I(X, 1/2) + I(X, -1/2) = 0$ in the above brute force analysis. Alternatively, this condition can be derived from a solvability condition.

Exercise 1 Derive the amplitude Eq. (16) by multiplying (13) by $\cos \pi y$ and then integrating the resulting equation from $y = -1/2$ to $y = 1/2$.

2 Localized Bulging of an Inflated Hyperelastic Tube

We now consider our first localization problem, inflation of a hyperelastic tube, for which bifurcation has a similar nature to the example in the previous section. Research on this problem dates back to Mallock (1885). For a review of the literature, we refer to Fu et al. (2008). For numerical illustrations, we shall use the Gent material model given by

$$W = -\frac{\mu}{2} J_m \ln(1 - \frac{I_1 - 3}{J_m}), \quad I_1 = \lambda_1^2 + \lambda_2^2 + \lambda_3^2, \tag{18}$$

and take $J_m = 97.2$, where μ is the ground state shear modulus, J_m is a constant characterising material extensibility, and λ_1, λ_2 and λ_3 are the three principal stretches.

There are two typical ways to inflate a rubber tube: by fixing the axial force or the total length of the tube. A non-zero fixed axial force is usually achieved by attaching a dead weight at one (closed) end of the tube, whereas in the case of fixed length the tube is usually subjected to a prestretch before inflation to avoid Euler buckling (the tube will elongate as soon as inflation starts). Figure 1 shows nine snapshots of the tube configuration in each case. Although the bulging processes in the two cases are seemingly similar, it will be shown that there are subtle differences. Our aim is to give the bulging process in each case a detailed and precise description under either mass control (corresponding to pumping air into the tube) or pressure control (corresponding to connecting the tube to a huge gas reservoir).

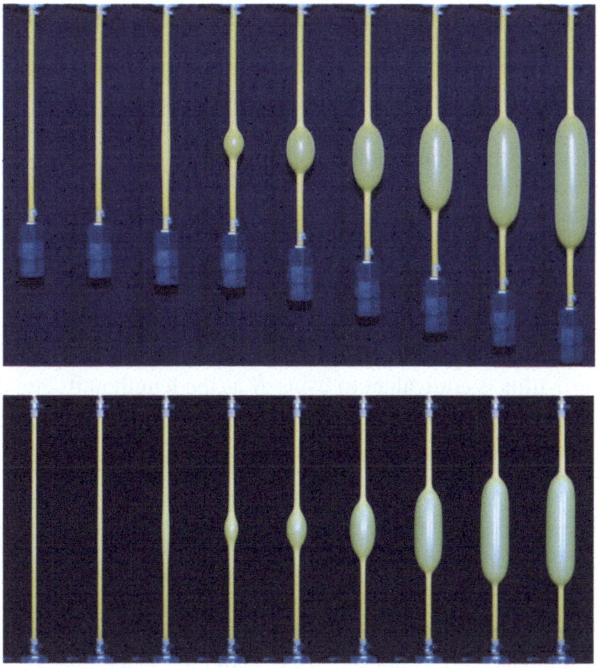

(a)

(b)

Fig. 1 Initiation, growth and propagation of a localized bulge when inflation is carried out with (a) N fixed or (b) total length fixed. From Wang et al. (2019)

2.1 Bifurcation Condition and Near-Critical Behaviour

To minimise algebra, we explain the methodology by assuming that the tube can be modelled as an incompressible membrane. Suppose that the tube has constant radius R, thickness H and length $2L$ before inflation, and the axisymmetric deformation is given by

$$r = r(Z), \quad \theta = \Theta, \quad z = z(Z)$$

in terms of cylindrical coordinates. Then the three principal stretches are given by

$$\lambda_1 = \frac{r}{R}, \quad \lambda_2 = \sqrt{r'^2 + z'^2}, \quad \lambda_3 = 1/(\lambda_1 \lambda_2).$$

The equilibrium equations can be obtained from the variational principle $\delta \mathcal{L} = 0$ with

$$\mathcal{L} = \int_{-L}^{L} H \tilde{W}(\lambda_1, \lambda_2) 2\pi R dZ - P \int_{-l}^{l} \pi r^2 dz - N \int_{-l}^{l} dz$$

$$= 2\pi R\mu H \int_{-L}^{L} (\bar{W} - \frac{1}{2}\bar{P}r^2 z' - \bar{N}z')dZ, \quad (19)$$

where $\tilde{W}(\lambda_1, \lambda_2) = W(\lambda_1, \lambda_2, (\lambda_1\lambda_2)^{-1})$ is the reduced strain energy per unit volume in the undeformed configuration, $l = z(L)$, N is the fixed axial force, P is the inflation pressure, and

$$\bar{W} = \frac{\tilde{W}}{\mu}, \quad \bar{P} = \frac{PR}{\mu H}, \quad \bar{N} = \frac{N}{2\pi\mu RH}.$$

In the case of fixed ends, the term involving $\bar{N}$ in (19) is set to zero. From now on we take $R = 1$ (this is equivalent to using R as the length unit) and drop the bars on W, P and N. It then follows that $\lambda_1 \equiv r$.

It can be shown that $\delta\mathcal{L} = 0$ yields

$$W_1 - (\frac{W_2}{\lambda_2}r')' - Prz' = 0, \quad \frac{W_2}{\lambda_2}z' - \frac{1}{2}P\lambda_1^2 = N, \quad |Z| < L, \quad (20)$$

where $W_1 = \partial W/\partial\lambda_1$, $W_2 = \partial W/\partial\lambda_2$. Both experiments and numerical simulations have shown that localized bulging is insensitive to the presence of ends (more precisely to the conditions imposed at the ends) when the tube is sufficiently long. Thus, we examine the situation where a bulge is localized around $Z = 0$ and the tube is *infinitely long*. We define

$$r_\infty \equiv \lambda_{1\infty} = \lim_{Z\to\pm\infty} \lambda_1(Z), \quad z_\infty = \lim_{Z\to\pm\infty} \lambda_2(Z).$$

The validity of the infinite length assumption will be examined later. To avoid introducing extra notations, we also use r_∞ and z_∞ to denote the values of λ_1 and λ_2 in uniform inflation. Thus, specialising the two equations in (20) to uniform inflation, we obtain

$$P = \frac{W_{1\infty}}{r_\infty z_\infty}, \quad N = W_{2\infty} - \frac{1}{2}Pr_\infty^2, \quad (21)$$

where $W_{1\infty} = W_1(r_\infty, z_\infty)$, etc.

When the Gent model (18) is adopted, we have

$$N = \frac{J_m\left(1 - 2z_\infty^4 r_\infty^2 + z_\infty^2 r_\infty^4\right)}{2\left(z_\infty^3 r_\infty^2\left(r_\infty^2 - J_m - 3\right) + z_\infty^5 r_\infty^2 + z_\infty\right)}. \quad (22)$$

Inflation can be carried out with either N or the total length fixed. Note that in the latter case, the tube would be stretched to a length given by $2z_\infty L$ and this length is then fixed in the subsequent inflation. Prior to bulge formation, fixing the total length is equivalent to fixing the axial stretch z_∞. When N is fixed, $(21)_2$ may be

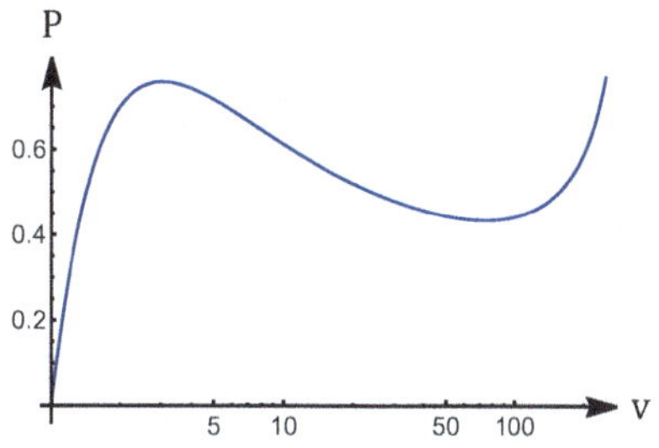

Fig. 2 Typical variation of P against volume ratio v in uniform inflation

used to express z_∞ in terms of r_∞. For instance, when the Gent model is used, $N = 0$ together with (22) can be solved to yield

$$z_\infty = \frac{1}{2}\sqrt{\frac{\sqrt{r_\infty^6 + 8}}{r_\infty} + r_\infty^2}. \tag{23}$$

For definiteness, in the following analysis for the case of fixed N, we assume that N is fixed to be zero, and z_∞ is given by (23). The volume ratio is given by $v = r_\infty^2 z_\infty$. The pressure versus v in uniform inflation with $N = 0$ for the Gent material model is shown in Fig. 2. It has the typical up-down-up shape. Note that in the case of fixed length, the pressure versus v would be monotonic (but bulging can still take place).

It can be shown with the use of $(20)_1$ and the differentiated form of $(20)_2$ that $W - \lambda_2 W_2 = C$ with C a constant. This and $(20)_2$ are the two integrals of the current problem which we rewrite here for convenience:

$$\frac{W_2}{\lambda_2} z' - \frac{1}{2} P \lambda_1^2 = N, \quad W - \lambda_2 W_2 = C. \tag{24}$$

The two constants N and C can be determined by evaluating the left hand sides of (24) at infinity, giving

$$\frac{W_2}{\lambda_2} z' - \frac{1}{2} P \lambda_1^2 = W_{2\infty} - \frac{1}{2} P r_\infty^2, \quad W - \lambda_2 W_2 = W_\infty - z_\infty W_{2\infty}. \tag{25}$$

On the other hand, on evaluating (25) at $Z = 0$ where we have $\lambda_2(0) = z'(0)$ due to $r'(0) = 0$, we obtain

$$W_2^{(0)} - \frac{1}{2} P r(0)^2 = W_{2\infty} - \frac{1}{2} P r_\infty^2, \tag{26}$$

$$W^{(0)} - \lambda_2(0) W_2^{(0)} = W_\infty - z_\infty W_{2\infty}, \tag{27}$$

where a superscript "(0)" signifies evaluation at $\lambda_1 = r(0)$, $\lambda_2 = \lambda_2(0)$. These are two algebraic equations for $r(0)$ and $\lambda_2(0)$ which can be solved to express $r(0)$ in

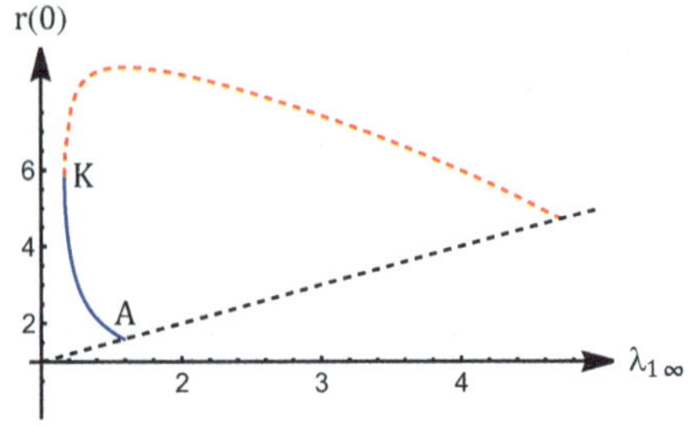

Fig. 3 Bifurcation diagram when inflation is carried out with $N = 0$ (i.e. closed ends without any extra dead weight). Points A and K correspond to the bifurcation point and Maxwell state, respectively

terms of r_∞, or equivalently the inflation pressure. Thus, under the infinite-length assumption, the bifurcation diagram can be determined without having to solve any differential equations. Corresponding to the Gent material model (18), this diagram takes the form shown in Fig. 3.

The bifurcation diagram gives us the following information. The black dashed line is the trivial branch where $r(0) = r(\infty) = \lambda_{1\infty}$, and inflation starts from the origin where $r(0) = \lambda_{1\infty} = 1$. When point A is reached, localized bulging would initiate and follow the blue solid line AK. Note that along this line, $r(\infty)$ decreases but $r(0)$ increases, which means that the tube is unloading at the two ends while the bulge grows at $Z = 0$. Point K is characterised by the property that $r''(0) = 0$ as well as $r'(0) = 0$, meaning that the state corresponding to K is an equilibrium state. Since the states at both $Z = 0$ and $Z = \infty$ are equilibrium states when point K is reached, the deformation in the tube is a "two-phase" or Maxwell state (Yin, 1977; Chater & Hutchinson, 1984). We next derive the bifurcation condition corresponding to point A explicitly.

Write the two integrals in (25) in the form

$$f(\lambda_1, \lambda_2) = 0, \qquad z' = g(\lambda_1, \lambda_2), \tag{28}$$

where f and g also depend on the uniform state (r_∞, z_∞) at infinity. Note that the trivial solution $\lambda_1 = r_\infty$, $\lambda_2 = z_\infty$ is always a solution.

For a bifurcated solution, write $r = \lambda_1 = r_\infty + y(Z)$. Equation $(28)_1$ then yields $\lambda_2 = \tilde{f}(y)$ and $(28)_2$ yields $z' = \tilde{g}(y)$, where $\tilde{f}(y)$ and $\tilde{g}(y)$ can be obtained with the use of *Mathematica*. From $\lambda_2 = \sqrt{r'^2 + z'^2} = \sqrt{y'^2 + z'^2}$ we then obtain

$$y'^2 = \lambda_2^2 - z'^2 = \omega(r_\infty)y^2 + \gamma(r_\infty)y^3 + \cdots,$$

or equivalently,

$$y'' = \omega(r_\infty)y + \frac{3}{2}\gamma(r_\infty)y^2 + \cdots, \tag{29}$$

where the expressions for the coefficients $\omega(r_\infty)$ and $\gamma(r_\infty)$ can easily be obtained using *Mathematica*. Note that r_∞ is taken to be the bifurcation parameter in the above equation with z_∞ either fixed or determined by (23).

Equation (29) is the fully nonlinear incremental equation, and its linearization is

$$y'' = \omega(r_\infty)y. \tag{30}$$

In the linear bifurcation analysis, we look for a normal mode solution of the form $y = Y\exp(ikZ)$ with Y a constant. On substituting this into (30), we obtain $\omega(r_\infty) + k^2 = 0$. Note that the solutions of (30) are sinusoidal when $\omega(r_\infty) < 0$ and exponential when $\omega(r_\infty) > 0$. Since the unstressed configuration with $r_\infty = 1$ should not be able to support a periodic mode (existence of such a mode would imply non-uniqueness of solutions), it follows that $\omega(1) > 0$. As inflation proceeds, we expect $\omega(r_\infty)$ to decrease, and the bifurcation condition $\omega(r_\infty) + k^2 = 0$ is first satisfied when $\omega(r_\infty) = 0$ with $k = 0$; cf. the sentence below (6). Thus, the bifurcation condition is

$$\omega(r_{\rm cr}) = 0. \tag{31}$$

The near-critical weakly nonlinear solution is governed by

$$y'' = \omega'_{\rm cr} \cdot (r_\infty - r_{\rm cr})\, y + \frac{3}{2}\gamma_{\rm cr} y^2, \qquad \omega'_{\rm cr} \equiv \omega'(r_{\rm cr}), \qquad \gamma_{\rm cr} \equiv \gamma(r_{\rm cr}), \tag{32}$$

which has a localized solution

$$y = -\frac{\omega'_{\rm cr}(r_\infty - r_{\rm cr})}{\gamma_{\rm cr}} {\rm sech}^2\left(\frac{1}{2}\sqrt{\omega'_{\rm cr}(r_\infty - r_{\rm cr})}\, Z\right). \tag{33}$$

Exercise 2 Use *Mathematica* to show that $\omega(r_\infty)$ is given by

$$\omega(r_\infty) = \frac{1}{r_\infty z_\infty W_2 W_{22}} \left\{ z_\infty^2 W_{22}(r_\infty W_{11} - W_1) - r_\infty (W_1 - z_\infty W_{12})^2 \right\}, \tag{34}$$

where $W_{22} = \partial^2 W / \partial \lambda_2^2$, etc. and are all evaluated at $\lambda_1 = r_\infty,\ \lambda_2 = z_\infty$.

Exercise 3 Show that the bifurcation condition $\omega(r_\infty) = 0$ with ω given by (34) is equivalent to

$$J(P, N) \equiv \frac{\partial P}{\partial r_\infty}\frac{\partial N}{\partial z_\infty} - \frac{\partial N}{\partial r_\infty}\frac{\partial P}{\partial z_\infty} = 0, \tag{35}$$

where P and N are given by (21). Show further that this condition reduces to

$$\frac{dP}{dr_\infty} = 0 \quad \text{when } N \text{ is fixed}, \tag{36}$$

and to

$$\frac{dN}{dz_\infty} = 0 \quad \text{when } P \text{ is fixed}. \tag{37}$$

More precisely, we have, e.g.,

$$\omega(r_\infty) = \frac{2r_\infty z_\infty^2 W_{22} + r_\infty^2 (W_1 - z_\infty W_{12})}{2W_2 W_{22}} \cdot \frac{dP}{dr_\infty} \quad \text{when } N \text{ is fixed.} \tag{38}$$

It can also be shown that the r_∞ in (35) and (36) can be replaced by the volume ratio v $(= r_\infty^2 z_\infty)$. Equation (36) shows that the bifurcation condition for localized bulging in the case of fixed axial force coincides with the condition for the limit point instability (Alexander, 1971). It will be shown later that Eq. (37) is the bifurcation condition for localized bulging in the case of fixed length, a result less intuitive than (36).

2.2 Graphical Illustration of the Bifurcation Condition

Corresponding to the Gent material model, the bifurcation condition (31) is plotted in Fig. 4a.

If inflation is carried out with $N = 0$, then the loading path is the solid blue line in Fig. 4a. Inflation starts at $(r_\infty, z_\infty) = (1, 1)$ and localized bulging would occur when point A is reached. Bulge evolution follows the path AK in Fig. 3. If inflation is carried out with fixed z_∞, then the loading path is a horizontal line in Fig. 4a. The red dashed line in Fig. 4a corresponds to $N(r_\infty, z_\infty) = 3.47$. Bifurcation becomes impossible if either N or λ_z is sufficiently large.

We may also plot the bifurcation condition in the (z_∞, P)-plane (Fig. 5a) or the (P, N)-plane (Fig. 5b). Figure 5a shows that localized bulging cannot take place if P is smaller than a threshold value or z_∞ is larger than a threshold value. These facts will feature in our discussions later.

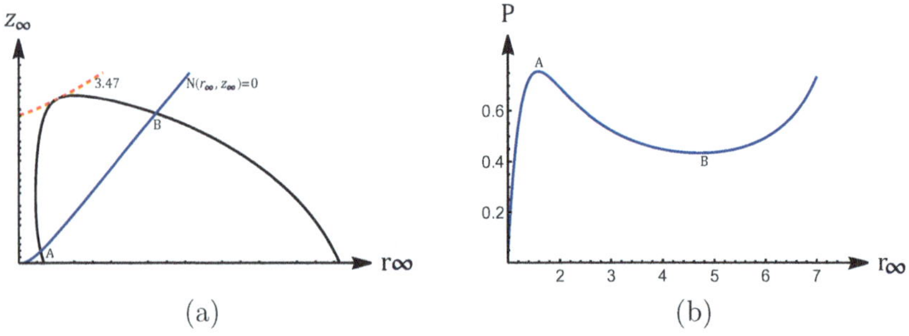

Fig. 4 (a) Bifurcation condition (black solid line) and the loading curve $N(r_\infty, z_\infty) = 0$. (b) Dependence of P on r_∞ in uniform inflation

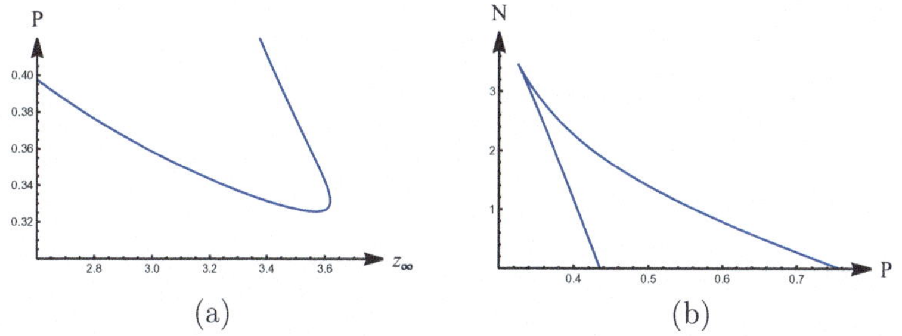

Fig. 5 Alternative representations of the bifurcation condition showing the fact that bifurcation is not possible if P is sufficiently small (left), or if N is sufficiently large (right)

2.3 Bulge Evolution—Fully Nonlinear Analysis

We first consider the case of fixed N. Figure 1a shows nine snapshots of a rubber tube that is inflated with fixed N (which is produced by a dead weight attached to the lower end). Figure 6a shows a typical set of Abaqus simulation results, and it is seen that both the inflation pressure P and bulge amplitude $r(0)$ tend to a constant value. The latter Maxwell state is determined by the equal area rule shown in Fig. 6b, as previously noted by Yin (1977), Chater and Hutchinson (1984) using analogies with phase transitions. This case has previously been well understood.

In contrast, the case of fixed tube length was not so well understood previously and there are very few analytical results. Figure 1b shows nine snapshots of a rubber tube that is inflated with fixed length. Although these snapshots are seemingly similar to their counterparts in Fig. 1a, the Abaqus simulation results shown in Fig. 7a are quite different from their counterparts in Fig. 6a. In particular, the pressure does not tend to a constant and the variation of P against $r(0)$ has an up-down-up behavior, as shown in Fig. 7b. To make sense of these numerical results, we present more numerical simulations in Fig. 8 by changing L at fixed λ_z $(= l/L)$ or changing λ_z at

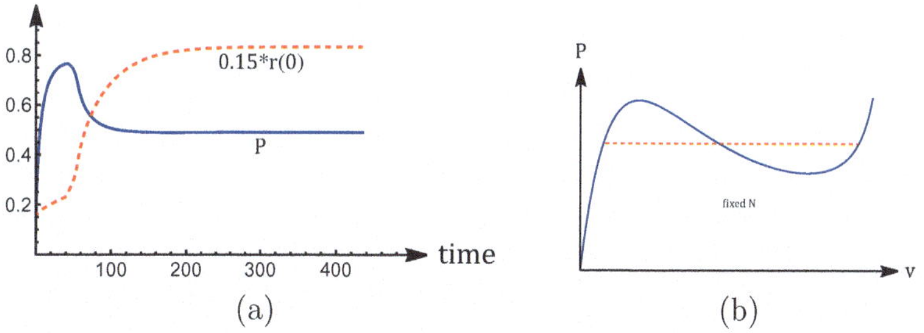

Fig. 6 Left: A typical set of Abaqus simulation results with $N = 0$. Pressure and bulging amplitude each approaches a constant. The propagation pressure is determined by the equal area rule in the P-v diagram on the right

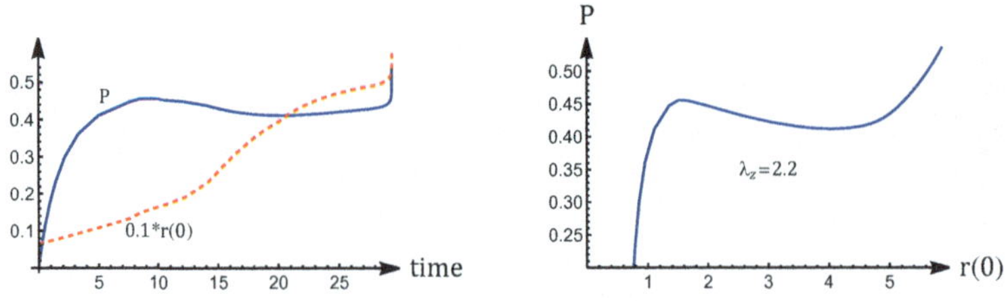

(a) (b)

Fig. 7 Abaqus simulation results for $L/R = 24$, $H/R = 0.4$ and $\lambda_z \equiv l/L = 2.2$. Left figure shows that in this case neither P nor $r(0)$ approaches a constant, but plotting P against $r(0)$, as in the right figure, shows that P has both a maximum and a minimum

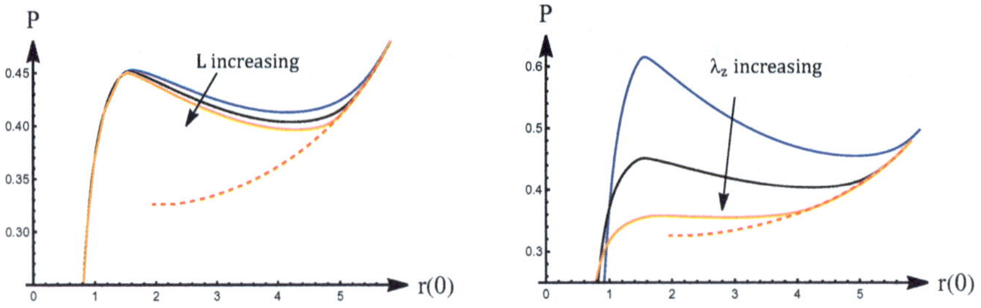

(a) $L = 15, 25, 30$ with $\lambda_z = 2.2$ (b) $\lambda_z = 1.5, 2.2, 3$ with $L = 25$.

Fig. 8 Abaqus simulation results for different values of L and fixed lengths reveal the existence of a master curve (the red dashed line)

fixed L. It is then seen that there exists a master curve (red dashed line in (a) and (b)) in each case. It turns out that this master curve can be described analytically, as we now show.

We first plot N against λ_z in Fig. 9 for different values of P. It is seen that for each $P > P_{\text{lower}} \approx 0.325$, the curve for N against λ_z has an up-down-up behavior. This P_{lower} corresponds to the pressure minimum in Fig. 5a. We may draw a horizontal line to cut the curve into equal areas, as was done in Fig. 10 for the case $P = 0.45$. This figure defines four values $\lambda_z^{(-)} < \lambda_z^{(\max)} < \lambda_z^{(\min)} < \lambda_z^{(+)}$. The variation of these four values with respect to P is displayed in Fig. 11. According to this figure, for each $P > P_{\text{lower}}$, there corresponds a value of $\lambda_z^{(-)}$, and hence a value of $\lambda_1^{(-)}$ via (23). Identifying the latter value with $r(0)$, and plotting P against $r(0)$, we obtain the master curves in Fig. 8a, b. This is further explained below.

Suppose first that the tube is inflated by increasing P with λ_z fixed at $\lambda_z = 3$ (thus on the left of $\lambda_z = 3.57$ where the pressure minimum is attained). The loading path is a vertical line in Fig. 11a. When P reaches the black line marked $\lambda_z^{(\max)}$, localized *bulging* initiates and if the tube length is fixed at $3 \times 2L$ while P is increased further, the tube will snap through to a Maxwell state with the "two phases" corre-

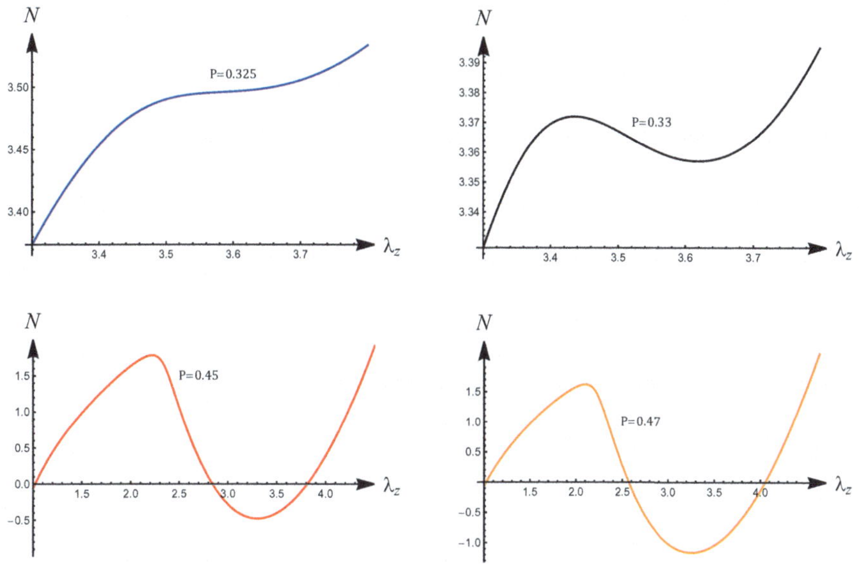

Fig. 9 N versus λ_z is monotonic when P is smaller than a certain value which we denote by P_{lower}, and the horizontal line (the Maxwell line) that cuts the curve into equal areas in the plot moves down as P increases

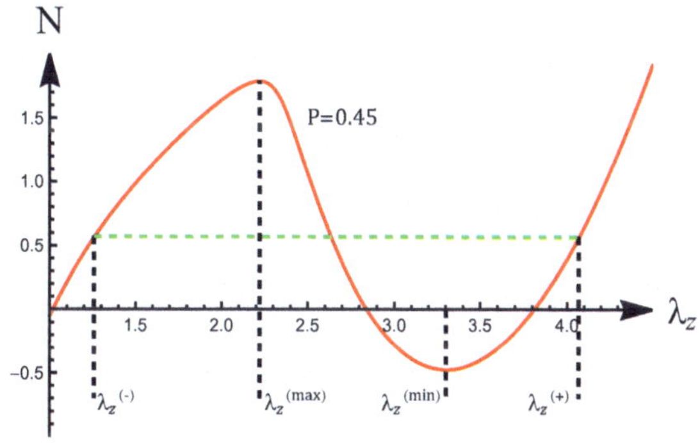

Fig. 10 The N versus λ_z when $P = 0.45$. The green line cuts the curve into equal areas. The green line would correspond to a negative N (buckling occurs) if $P > P_{\text{upper}}$. Each value of P in $(P_{\text{lower}}, P_{\text{upper}})$ corresponds to a Maxwell state

sponding to $\lambda_z^{(-)}$ and $\lambda_z^{(+)}$, respectively. Further evolution of the Maxwell state as P is increased follows the two branches in blue. If, on the other hand, the tube is inflated by increasing P with λ_z fixed at $\lambda_z = 3.6$, which is larger than 3.57 but less than the value corresponding to the vertical dashed line in Fig. 11b, localized *necking* will occur when the black curve is reached and the tube will again snap through to a Maxwell state with the "two phases" taking the values $\lambda_z^{(-)}$ and $\lambda_z^{(+)}$. In the special case when inflation is carried out with λ_z fixed at $\lambda_z = 3.57$ (where P attains its minimum), an exceptionally super-critical bifurcation takes place when the minimum of P is reached. The tube would evolve smoothly into a Maxwell state

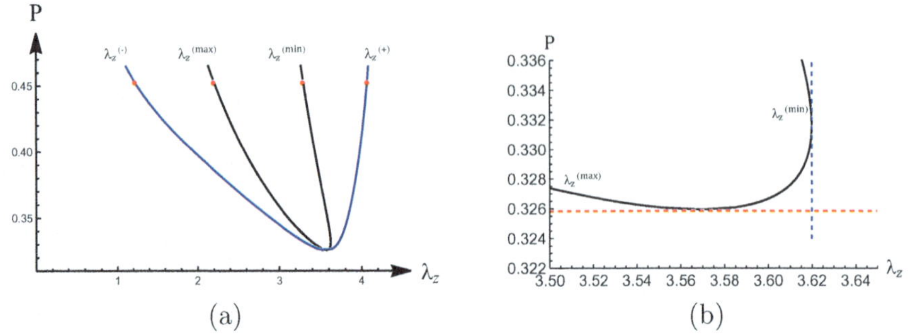

Fig. 11 (a) The bifurcation values and Maxwell values of λ_z as P increases above its minimum. (b) Blow-up of the black curve in (a) near the minimum of P at (3.57, 0.326)

without a snap-through. Plotting the corresponding variation of P versus $r(0)$ then gives the master curve in Fig. 8b. For more details, see Guo et al. (2022).

To summarize, when inflation is carried out with fixed axial force N, we may refer to Fig. 6 to deduce what will happen. If inflation is under volume control, pressure is allowed to decrease and Fig. 6a depicts how P and $r(0)$ vary with respect to time. If, on the other hand, inflation is under pressure control (the tube would then need to be connected to a huge reservoir of gas) and P is increased beyond its maximum, the tube will snap through to a uniform state corresponding to the right branch in Fig. 6b.

In the other case when inflation is carried out with fixed tube length, we may refer to Fig. 7b to deduce what will happen. If inflation is under volume control, pressure is allowed to decrease and Fig. 7b depicts how P varies with respect to the bulge amplitude $r(0)$ and a Maxwell state will be approached asymptotically as inflation continues. If inflation is under pressure control and P is increased beyond its maximum, the tube will snap through to a Maxwell state corresponding to the right branch in Fig. 7b, as discussed above in the context of Fig. 11. This completes our description, under the membrane assumption, of the entire inflation process whether the inflation is carried out with fixed N or fixed tube length. When the membrane assumption is removed and a tube of arbitrary wall thickness is considered, a 1D gradient model can be derived based on the exact theory of nonlinear elasticity and the same degree of understanding can still be achieved; see Yu and Fu (2023) for details.

2.4 A 1D Gradient Model Under the Membrane Assumption

In this subsection, we introduce the 1D gradient model of Lestringant and Audoly (2018) and assess its validity by comparing its predictions with fully nonlinear Abaqus simulations. Once its accuracy is assured, we will use it in the next subsection to assess the validity of our infinite length assumption.

The essence of the 1D gradient model is to assume that the spatial variation is small/slow and to only keep its leading order contribution. This assumption is obviously valid in the weakly nonlinear regime, where the variation is indeed slow, and also in the propagation stage where the deformation is piecewise constant except over the thin transition layer that connects the two Maxwell states. It turns out that the 1D model gives sufficiently accurate results for the entire inflation process.

Under the assumption that $|r'| \ll 1$, we expect $z' = g(r)$ to leading order, where $g(r)$ is the solution of

$$W_2(r, g(r)) - \frac{1}{2}Pr^2 - N = 0. \tag{39}$$

Note that (39) is simply $(21)_2$ with r_∞ and z_∞ replaced by $r(Z)$ and $g(r)$, respectively. To obtain an asymptotically self-consistent solution, we add correction terms and write

$$z' = g(r) + k(r)r'^2 + \cdots, \tag{40}$$

where the coefficient $k(r)$ is to be determined. On substituting this expansion into $\lambda_2 = \sqrt{r'^2 + z'^2}$ and the resulting expression into the integrand of the final integral in (19), we obtain

$$W(r, \lambda_2) - \frac{1}{2}Pr^2 z' - Nz' = G(r, r') + \frac{1}{2}W_{22}Q(r)r'^4 + O(r'^6), \tag{41}$$

where

$$G(r, r') = W(r, g(r)) - \frac{1}{2}Pr^2 g(r) - Ng(r) + W_2(r, g(r))\frac{r'^2}{2\,g(r)}, \tag{42}$$

$$Q(r) = \left(k - \frac{W_2 - gW_{22}}{2g^2 W_{22}}\right)^2 + \frac{gW_2 W_{22} - W_2^2}{4\,g^4 W_{22}^2}. \tag{43}$$

In (43), W_2 and W_{22} denote $W_2(r, g(r))$ and $W_{22}(r, g(r))$, respectively. We now minimize the right hand side of (41) with respect to the disposable function $k(r)$. The first term that contains $k(r)$ is the $O(r'^4)$ term. Minimizing this term with respect to $k(r)$, we obtain

$$k(r) = \frac{W_2 - gW_{22}}{2\,g^2 W_{22}}. \tag{44}$$

Not surprisingly, this same expression can also be obtained by substituting (40) and $\lambda_2 = \sqrt{r'^2 + z'^2}$ into $(24)_1$, expanding the resulting expression in terms of r', and then equating the coefficient of r'^2 to zero (the $O(1)$ term being zero due to (39)). However, the methodology adopted above, i.e. choosing the leading order solution to satisfy one of the two equilibrium equations as if the solution were uniform and then adding terms to make the solution asymptotically self-consistent through energy minimisation, can be extended more easily to the case of a hyperelastic tube of arbitrary wall thickness.

A 1D gradient model that takes the leading gradient term into account is given by $\delta\tilde{\mathcal{L}} = 0$ with

$$\tilde{\mathcal{L}} = \int_{-L}^{L} G(r, r')dZ. \tag{45}$$

The associated Euler-Lagrange equation is given by

$$\frac{\partial G}{\partial r} - \frac{d}{dZ}\left(\frac{\partial G}{\partial r'}\right) = 0,$$

that is,

$$\frac{W_2}{g} r'' = W_1 - Prg - (W_{12} + W_{22} g g' - W_2 g') \frac{r'^2}{2g^2}. \tag{46}$$

The $g'(r)$ in the above equation can be determined by differentiating the identity (39) with respect to r, and is given by

$$g'(r) = \frac{Pr - W_{12}}{W_{22}}. \tag{47}$$

It has been demonstrated by Lestringant & Audoly (2018) that although the 1D model (46) subject to the natural boundary conditions $r'(0) = r'(L) = 0$ is only meant to be valid if $|r'|$ is sufficiently small, it can accurately describe the entire inflation process. This is also demonstrated in the following subsection where its predictions are compared with Abaqus simulations.

2.5 Evaluation of the Infinite Length Assumption

In Sect. 2.2 we assumed that the tube length is infinite. We now show that the theory based on this assumption gives excellent result even when L, which has been scaled by R, is only moderately large.

We impose the natural boundary conditions $r'(\pm L) = 0$ at the two ends, and remark that more general end conditions (e.g. fixed radius) do not in general allow uniform inflations. This makes a strong case for the alternative view of assuming the tube to be infinitely long and treating the presence of ends as an imperfection. In the latter view, the precise forms of the end conditions become immaterial.

For a bulging solution, we may assume that the profile is symmetric with respect to $Z = 0$, and hence $r'(0) = 0$. Thus, we may impose the boundary conditions

$$r'(0) = 0, \quad r'(L) = 0. \tag{48}$$

The second order nonlinear ODE (46) subject to the BCs (48) can easily be solved using the central finite difference scheme. We discretize the interval $[0, L]$ into n

intervals with node points $Z_i = iL/n$ $(i = 0, 1, \ldots, n)$. At the i-th node, the first- and second-order derivatives are replaced by

$$r' = \frac{r_{i+1} - r_{i-1}}{2h}, \quad r'' = \frac{r_{i+1} - 2r_i + r_{i-1}}{h^2}, \quad r_i \equiv r(Z_i) \quad h = \frac{L}{n}. \tag{49}$$

Applying this scheme at the node points $Z_0, Z_1, \ldots, Z_n$, we obtain $n+1$ nonlinear homogeneous equations for the $n+1$ unknowns $r_0, r_1, \ldots, r_n$ with r_{-1} and r_{n+1} eliminated using $r_{-1} = r_1$ and $r_{n+1} = r_{n-1}$ due to (48). This scheme works very well as long as a good initial guess is provided. The initial guess is provided by the weakly nonlinear theory which we derive next.

We use r_∞ as the bifurcation parameter and assume that z_∞ is determined from (23). The corresponding inflation pressure is then computed using $(21)_1$. We write

$$r_\infty = r_{\mathrm{cr1}} + \epsilon^2 r_0, \tag{50}$$

where r_{cr1} is the bifurcation value of r_∞, ϵ is a small positive parameter characterising the deviation of r_∞ from r_{cr1}, and r_0 is a constant. We look for an asymptotic solution of the form

$$r(Z) = r_\infty + \epsilon r_1(Z) + \epsilon^2 r_2(Z) + \epsilon^3 r_3(Z) + \cdots, \tag{51}$$

$$z(Z) = z_\infty + \epsilon z_1(Z) + \epsilon^2 z_2(Z) + \epsilon^3 z_3(Z) + \cdots, \tag{52}$$

where the various functions on the right hand sides are to be determined as successive orders. We assume for the moment that all the derivatives of these functions are of $O(1)$. By substituting (51) and (52) into the two equilibrium equations (20) and the two boundary conditions (48), and then equating the coefficients of powers of $\epsilon, \epsilon^2, \epsilon^3$, we obtain three boundary value problems. At the i-th order $(i = 1, 2, 3)$, the unknown $z_i(Z)$ can be expressed in terms of $r_i(Z)$ by solving a linear equation (obtained from $(20)_2$) and the governing equation consists of a single second order ODE for $r_i(Z)$. It is found that $r_1(Z)$ satisfies

$$r_1''(Z) - \omega(r_{\mathrm{cr1}}) r_1(Z) = 0, \quad r_1'(0) = r_1'(L) = 0. \tag{53}$$

We observe that $(53)_1$ is of the same form as (30). It is then found that

$$\omega(r_{\mathrm{cr1}}) + k_{\mathrm{cr}}^2 = 0, \quad r_1(Z) = A\cos(k_{\mathrm{cr}} Z), \quad k_{\mathrm{cr}} = \frac{\pi}{L}, \tag{54}$$

where the amplitude A is to be determined at third order by a solvability condition. Since $\omega(\lambda_\theta)$ is proportional to dP/dv, see (38), the left hand side of $(54)_1$ is positive before P reaches its maximum, and so $(54)_1$ can only be satisfied after P has reached its maximum.

We note that the eigenvalue problem (53) also admits a solution in which r_1 is a constant provided $\omega(r_{\mathrm{cr1}}) = 0$. This solution is dismissed for the moment because it corresponds to uniformly inflated solutions. However, when L becomes large, the solution $(54)_2$ becomes invalid since it is slowly varying and hence contradicts our earlier assumption that $r_1'(Z), r_1''(Z)$ etc. are all of $O(1)$. In other words, a large L allows r_1 to vary slowly so that to leading order r_1 is like a constant, and as a result $\omega(r_{\rm cr1}) = 0$ becomes the bifurcation condition. This case will be discussed shortly.

The second order solution is given by

$$r_2(Z) = A^2\left(q_0 + q_1 \cos(2k_{\rm cr}Z)\right), \tag{55}$$

where the explicit expression for q_0 and q_1 are not written out here for brevity. At third order, elimination of secular terms leads to the solvability condition

$$r_0 A - c_1 A^3 = 0, \quad \Longrightarrow \quad \epsilon A = \sqrt{\frac{r_\infty - r_{\rm cr1}}{c_1}}, \tag{56}$$

where the explicit expressions for the constant c_1 is not written out here for brevity. It is found that c_1 is positive for all values of L. Thus, the bifurcation solution exists only for $r_0 > 0$. Since P is a monotonically decreasing function of r_∞ near the bifurcation point, $r_0 > 0$ means that $P < P_{\rm cr}$. Thus, the bifurcation is subcritical, and the two-term near-critical bifurcated solution is given by

$$r(Z) = (\epsilon A)\cos(k_{\rm cr}Z) + (\epsilon A)^2\left\{q_0 + q_1 \cos(2k_{\rm cr}Z)\right\}. \tag{57}$$

This can be used as the initial guess in the finite difference scheme in order to trace the solutions into the fully nonlinear regime.

When L becomes large, the critical wavenumber $k_{\rm cr}$ becomes small. More precisely, the solution (57) suggests that when L is of order $1/\sqrt{\epsilon}$, the solution depends on Z through the far-distance variable $s = \sqrt{\epsilon}Z$. An alternative argument leading to the same conclusion is that the eigenvalue problem (53) admits the non-trivial solution $r_1(Z) =$ const with $k_{\rm cr} = 0$. Thus, the problem belongs to the type discussed in Sect. 1.2. In this case, we write $r_\infty = r_{\rm cr} + \epsilon \hat{r}_0$ and look for an asymptotic solution of the form

$$r(Z) = r_\infty + \epsilon \hat{y}(s) + \cdots. \tag{58}$$

The asymptotic expansion for $z'(Z)$ may be obtained from $(20)_2$ by simple Taylor expansion. On substituting these two expressions into $(20)_1$ and equating the coefficients of ϵ and ϵ^2, we obtain the bifurcation condition (31) and the amplitude equation

$$\hat{y}'' = \omega'_{\rm cr}\hat{r}_0\hat{y} + \frac{3}{2}\gamma_{\rm cr}\hat{y}^2. \tag{59}$$

Note that (59) is simply a scaled form of (32), but here the boundary conditions are

$$\hat{y}'(0) = \hat{y}'(\sqrt{\epsilon}L) = 0. \tag{60}$$

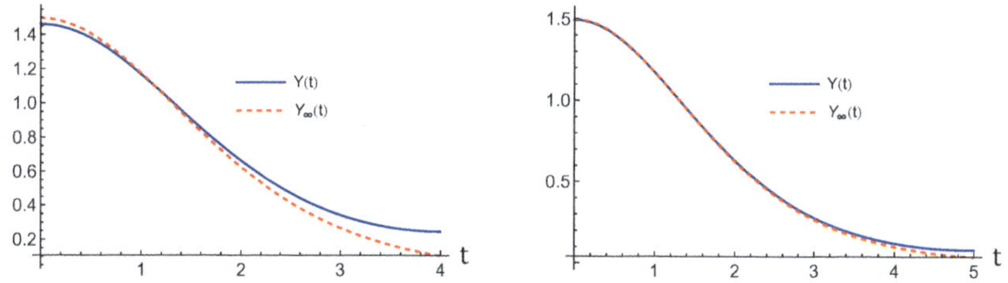

(a) $\tilde{L} = 4$ (b) $\tilde{L} = 5$

Fig. 12 Comparison of $Y(t)$ with $Y_\infty(t)$ when $\tilde{L} = 4$ or 5

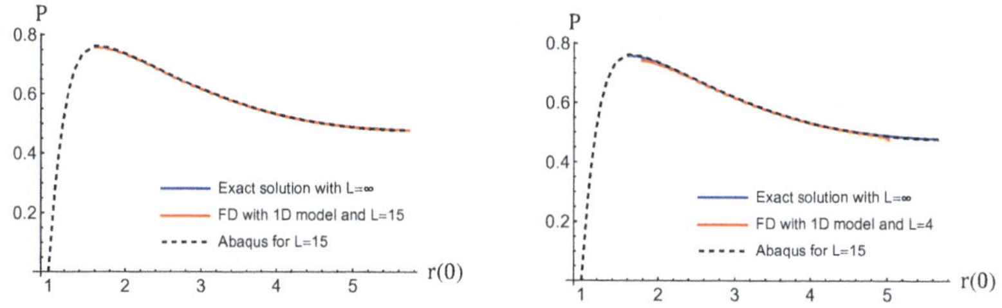

(a) (b)

Fig. 13 Dependence of P on $r(0)$ as the bulge evolves, comparing the FD solution (based on the 1D model) with the Abaqus simulation result for (a) $L = 15$ and (b) $L = 4$. Also displayed in both figures is the exact result based on the infinite-length theory

With the substitution

$$\hat{y}(s) = -\frac{2\omega'_{\rm cr}\hat{r}_0}{3\gamma_{\rm cr}}Y(t), \quad t = \sqrt{\omega'_{\rm cr}\hat{r}_0}\, s,$$

we obtain

$$Y'' = Y - Y^2, \quad Y'(0) = Y'(\tilde{L}) = 0, \tag{61}$$

where $\tilde{L} = \sqrt{(r_\infty - r_{\rm cr})\omega'_{\rm cr}}L$. In the limit $\tilde{L} \to \infty$, the unique solution of the above BVP with $Y(0) > Y(\tilde{L})$ tends to $Y_\infty(t) = (3/2)\text{sech}^2(t/2)$, a scaled form of the localized solution (33). This solution of $(61)_1$ subject to $(61)_{2,3}$ can be written in terms of an elliptic function or can be obtained simply by a shooting procedure. Figure 12 shows that when $\tilde{L} = 5$, $Y(t)$ is well approximated by $Y_\infty(t)$ except near the tube end. This solution can equally be used as the initial guess in the above-mentioned finite difference scheme.

Figure 13a, b compare the 1D theory with Abaqus simulations for the dependence of P on $r(0)$ as the bulge evolves. It is seen that the 1D theory gives accurate results for the entire inflation process, from bulge initiation to bulge propagation. Also, it is seen that the only major difference between the results for $L = 4$ and $L = \infty$ is in the bulge initiation point, and the exact theory based on the infinite length assumption provides a very good approximation for tubes of sufficiently large lengths.

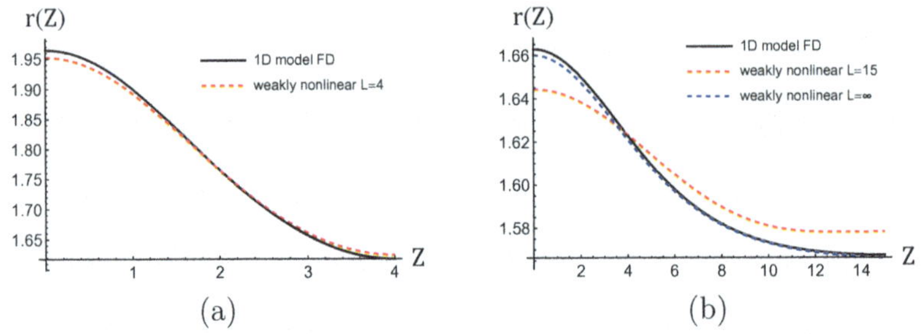

Fig. 14 Comparison of FD solution (based on the 1D model) with the weakly nonlinear solution (57) when $r_\infty - r_{\rm cr1} = 0.05$ and (a) $L = 4$ or (b) $L = 15$. The blue dashed line in (b) corresponds to the weakly nonlinear solution (33) based on the infinite length assumption

Figure 14a, b show profiles of $r(Z)$ in the near-critical regime. It is seen that whereas the two-term weakly nonlinear solution (57) provides a good approximation for the exact solution (which is very well approximated by the 1D theory) when $L = 4$, it fails when $L = 15$. For the latter case, it is outperformed by the solution based on (59) and (60) (or the weakly nonlinear theory based on the infinite length assumption).

2.6 Tubes of Finite Wall Thickness

For a hyperelastic tube of finite wall thickness, localized bulging can be investigated using the same strategy as in the previous subsections although algebraically more involved. In particular, the bifurcation condition was derived by Fu et al. (2016), a weakly nonlinear analysis was conducted by Ye et al. (2020), and a 1D reduced model was presented by Yu and Fu (2023) with the aid of which the entire bulging process can be investigated using a simple finite difference scheme in the same manner as in the previous subsection.

3 Necking of a Solid Cylinder Under Surface Tension

When a neo-Hookean tube is axially compressed, it may lose stability and adopt a configuration in which the outer radius varies periodically in the axial direction; see Wilkes (1955). The principal stretch λ is a function of k, the axial wavenumber, and it has a unique maximum at a finite wavenumber (the critical wavenumber). In the limit when the inner radius tends to zero (the case of a solid cylinder), the critical wavenumber tends to infinity, and the bifurcation mode is localized near the curved surface like a surface wave mode (assuming that Euler buckling does not occur). Localized necking or bulging will not occur.

However, when a solid cylinder is subjected to *surface tension* and axial extension, it will suffer localized necking or bulging, depending on the axial stretch ratio. The surface tension may be applied in two different ways, either with fixed axial force or fixed length, rather like the situations in the inflation problem. We note that surface tension tends to shrink the cylinder in the axial direction (imagine wrapping the soft cylinder by a thin film in tension). Thus, if the total length of the cylinder is fixed, it will become more and more stretched as surface tension is increased. We now show that the close analogy with the tube inflation problem can be used to understand the current problem with little extra effort.

Consider a hyperelastic solid cylinder that is defined in terms of cylindrical coordinates by $R \le B, 0 \le \Theta \le 2\pi$ and $|Z| < L$, where B is the undeformed radius and L is the half-length. The homogeneous deformation is given by

$$r = \frac{1}{\sqrt{\lambda}} R, \quad \theta = \Theta, \quad z = \lambda Z, \tag{62}$$

where (r, θ, z) are cylindrical coordinates in the deformed configuration. The associated deformation gradient is given by

$$\mathbf{F} = \frac{1}{\sqrt{\lambda}} \boldsymbol{e}_r \otimes \boldsymbol{E}_R + \frac{1}{\sqrt{\lambda}} \boldsymbol{e}_\theta \otimes \boldsymbol{E}_\Theta + \lambda \boldsymbol{e}_z \otimes \boldsymbol{E}_Z, \tag{63}$$

where λ is the *constant* stretch in the z-direction. The cylinder is assumed to be incompressible with its constitutive behavior described by a generalized neo-Hookean strain-energy function of the form

$$W = w(I_1), \quad I_1 = \mathrm{tr}\, \mathbf{F}\mathbf{F}^{\mathrm{T}}. \tag{64}$$

The Cauchy stress is given by

$$\boldsymbol{\sigma} = 2w'(I_1)\mathbf{F}\mathbf{F}^{\mathrm{T}} - pI, \tag{65}$$

where p is the Lagrange multiplier associated with the incompressibility constraint. For the uniform deformation, application of the boundary condition $\sigma_{rr} = -\gamma/b$ at the outer surface $r = b$ yields

$$p = \frac{2w_d}{\lambda} + \frac{\gamma}{b}, \tag{66}$$

where w_d denotes $w'(I_1)$ evaluated at $I_1 = \lambda^2 + 2\lambda^{-2}$. The Cauchy stress in the axial direction is then given by

$$\sigma_{zz} = 2(\lambda^2 - \lambda^{-1})w_d - \frac{\gamma}{b}. \tag{67}$$

The resultant axial force N is computed according to $N = \pi b^2 \sigma_{zz} + 2\pi b\gamma$ and is given by

$$N = \pi \left(2(\lambda - \lambda^{-2}) w_d + \frac{\gamma}{\sqrt{\lambda}} \right). \tag{68}$$

If N is fixed, the surface tension γ becomes a function of the axial stretch λ. If on the other hand γ is fixed, N becomes a function of λ. Based on our experience with the inflation problem, we may anticipate that the bifurcation condition is given by $d\gamma/d\lambda = 0$ if N is specified, or by $dN/d\lambda = 0$ if γ is specified. These two conditions yield

$$N_{cr} = 2\pi \frac{3\lambda^4 w_d + 3\lambda w_d + 4\lambda^6 w_{dd} - 8\lambda^3 w_{dd} + 4 w_{dd}}{\lambda^3}, \tag{69}$$

and

$$\gamma_{cr} = \frac{4\lambda\left(\lambda^3 + 2\right) w_d + 8\left(\lambda^3 - 1\right)^2 w_{\mathrm{dd}}}{\lambda^{5/2}}, \tag{70}$$

respectively. It can be shown that substituting (68) into (69) (with N and γ replaced by N_{cr} and γ_{cr}, respectively) would lead to (70). Thus, (69) and (70) are equivalent bifurcation conditions.

A linear bifurcation analysis can be conducted to show that if the wavenumber in the axial direction is k, then a bifurcation condition of the form $f(\gamma, \lambda, k) = 0$ can be determined analytically; see, e.g., Taffetani and Ciarletta (2015). The dependence of γ on k for each fixed λ, or the dependence of λ on k for each fixed γ, can be obtained numerically. A typical set of results is given in Fig. 15. It is seen that the minimum of λ or γ is attained at $k = 0$.

It can then be concluded that localized bulging or necking will occur in the current problem. A typical plot of the bifurcation condition (70) is given in Fig. 16. This plot may be compared with Fig. 5a (or the blow-up in Fig. 11b). There is a close analogy between the current problem and the tube inflation problem discussed in Sect. 3:

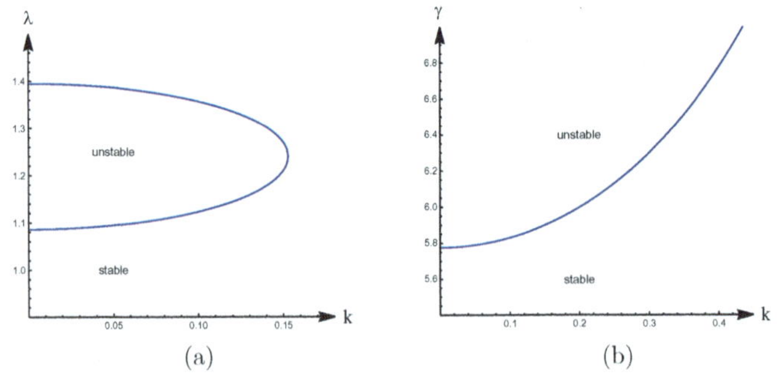

Fig. 15 Bifurcation condition showing the dependence of (a) λ on k when $\gamma = 5.8$, and (b) γ on k when $\lambda = 1.1$. Both curves correspond to the Gent material model with $J_m = 97.2$

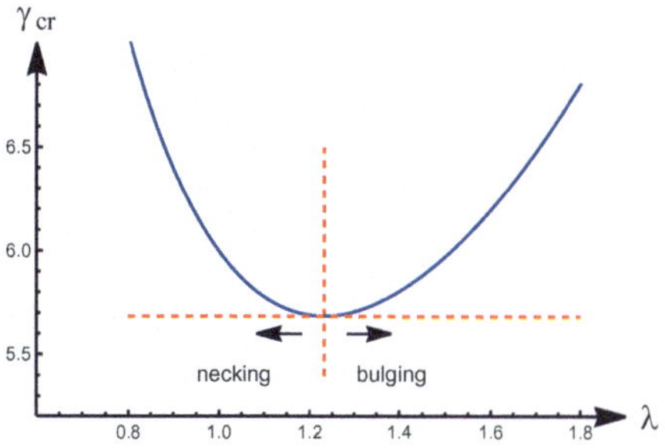

Fig. 16 Dependence of γ_{cr} on λ corresponding to the Gent material model with $J_m = 97.2$

the case of varying γ with fixed length here is very similar to the case of varying pressure with fixed length there, whereas the case of varying λ with fixed γ is similar to the case of varying λ with fixed pressure P. The following conclusions can then be drawn without further analysis (Fig. 17):

1 If loading is carried out by varying γ at a fixed λ, then the cylinder will snap through to a Maxwell state when the bifurcation point is reached. The Maxwell state is determined by the equal area rule in the N vs λ diagram at each value of γ. Thus, the Maxwell state is γ-dependent, and the dependence of the two radii on γ can be determined semi-analytically.
2 If loading is carried out by varying λ at a fixed γ, the axial force will start to decrease when the bifurcation point is reached and then the deformation rapidly approaches the Maxwell state associated with this particular γ. The Maxwell state is again determined by the equal area rule in the N vs λ diagram.

Details of the above analysis are given in Fu et al. (2021); see also Xuan and Biggins (2017) who focused on the behaviour in the fully nonlinear regime. Analogous analyses for a soft tube subject to surface tension have been carried out by Emery and Fu (2021a, b).

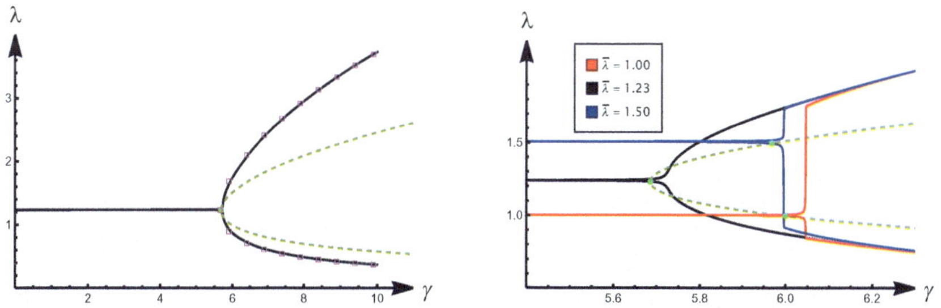

(a) (b)

Fig. 17 (a) Bifurcation curve (dashed green curve) and Maxwell states (black line) with the squares denoting numerical simulation results. (b) Numerical simulation results for loading at three different values of the average stretch $\bar{\lambda}$ (which is the same as the λ_z in the text). The cyan dots are the bifurcation points predicted by the theory. From Fu et al. (2021)

4 Axisymmetric Necking in a Circular Hyperelastic Plate Under Equibiaxial Stretching

In this section, we consider axisymmetric necking in a hyperelastic plate under all-round stretching; see Fig. 18. Note that all-round stretching in the plane is equivalent to equibiaxial stretching. We may start with uniform *biaxial* stretching with the two in-plane nominal stresses and stretches denoted by S_1, S_2 and λ_1, λ_2, respectively (think of a large square plate). The two nominal stresses are functions of the two principal stretches. It is natural to compute the Jacobian determinant of S_1 and S_2 and ask what happens when this Jacobian determinant vanishes, especially when evaluated $\lambda_1 = \lambda_2$ (the equibiaxial case). It turns out that the vanishing of this determinant does not correspond to localisation; the condition for localized necking is given by a different bifurcation condition which we now derive.

4.1 Governing Equations and the Primary Solution

We first summarise the governing equations for general buckling analysis under the framework of nonlinear elasticity. In general terms, we consider an elastic body which has an undeformed configuration B_0 in a three-dimensional Euclidean point space. This elastic body is then subjected to a finite deformation $\boldsymbol{x} = \chi(\boldsymbol{X})$ and the resulting configuration is denoted by B_e, where $\boldsymbol{X}$ and $\boldsymbol{x}$ are the position vectors of a representative material particle in B_0 and B_e, respectively. In order to determine whether other configurations are possible, we first assume that another (buckling) solution is indeed possible and we denote it by $\tilde{\boldsymbol{x}} = \tilde{\chi}(\boldsymbol{x})$, where $\tilde{\boldsymbol{x}}$ is the position vector, in the buckled configuration B_t, of a material particle whose position vector in B_e is $\boldsymbol{x}$. Since $\tilde{\boldsymbol{x}}$ and $\boldsymbol{x}$ satisfy the same governing equations and boundary conditions,

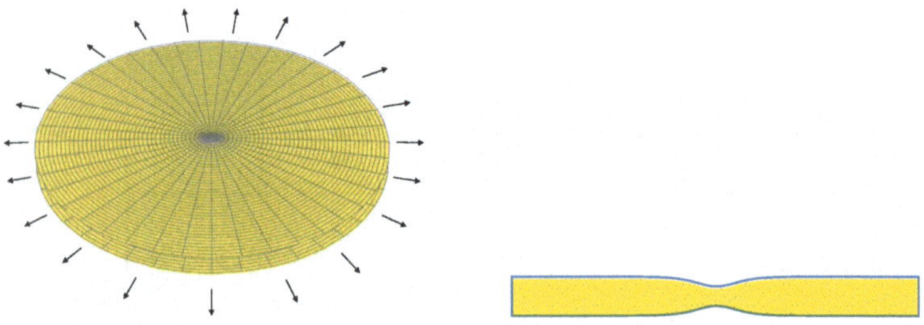

(a) (b)

Fig. 18 (a) Axisymmetric necking of a circular plate under all-round stretching. (b) Side view of the plate along a diameter

the incremental displacement $\boldsymbol{u}$, defined by $\boldsymbol{u} = \tilde{\boldsymbol{x}} - \boldsymbol{x}$, satisfies a nonlinear boundary value problem which always admits the trivial solution $\boldsymbol{u} = \boldsymbol{0}$.

We denote the deformation gradients from B_0 to B_e and from B_0 to B_t by $\boldsymbol{F}$ and $\tilde{\boldsymbol{F}}$, respectively. It follows from $\tilde{\boldsymbol{F}} d\boldsymbol{X} = d\tilde{\boldsymbol{x}} = d\boldsymbol{x} + d\boldsymbol{u} = d\boldsymbol{x} + (\text{grad}\,\boldsymbol{u})d\boldsymbol{x} = (I + \text{grad}\,\boldsymbol{u})\boldsymbol{F} d\boldsymbol{X}$ that

$$\tilde{\boldsymbol{F}} = (I + \boldsymbol{\eta})\boldsymbol{F}, \tag{71}$$

where $\boldsymbol{\eta} = \text{grad}\,\boldsymbol{u}$ and the gradient operator grad is with respect to $\boldsymbol{x}$. The dead-load traction boundary condition corresponds to $(\tilde{\boldsymbol{S}}^T - \boldsymbol{S}^T)\boldsymbol{N} = \boldsymbol{0}$, where $\tilde{\boldsymbol{S}}$ and $\boldsymbol{S}$ are the nominal stresses in B_t and B_e, respectively, and $\boldsymbol{N}$ is the normal to a material surface in B_0 where dead-load tractions are prescribed. With the use of Nanson's formula, the dead-load traction boundary condition may also be written as $(\tilde{\boldsymbol{S}}^T - \boldsymbol{S}^T)\boldsymbol{F}^T\boldsymbol{n} = 0$, where $\boldsymbol{n}$ is the normal to the same material surface in B_e. This motivates the introduction of an incremental stress tensor $\boldsymbol{\chi}$ through

$$\boldsymbol{\chi} = J^{-1}(\tilde{\boldsymbol{S}}^T - \boldsymbol{S}^T)\boldsymbol{F}^T, \tag{72}$$

where $J = \det(\boldsymbol{F})$ is inserted to enable application of the useful property $\text{div}(J^{-1}\boldsymbol{F}) = \boldsymbol{0}$ (i.e. $(J^{-1}\boldsymbol{F}_{jA})_{,j} = 0$ in rectangular coordinates). The dead-load boundary condition then reduces to $\boldsymbol{\chi}\boldsymbol{n} = \boldsymbol{0}$ and the two equilibrium equations $\text{Div}\,\tilde{\boldsymbol{S}} = 0$ and $\text{Div}\,\boldsymbol{S} = 0$ imply the incremental equilibrium equation

$$\text{div}\,\boldsymbol{\chi}^T = \boldsymbol{0}. \tag{73}$$

Assuming that the displacement gradient $\boldsymbol{\eta}$ has a small amplitude, we may expand the incompressibility condition $\det \tilde{\boldsymbol{F}} = 1$ and the right hand side of (72) around $\tilde{\boldsymbol{F}} = \boldsymbol{F}$ to obtain

$$\eta_{ii} = \frac{1}{2}\eta_{mn}\eta_{nm} - \frac{1}{2}(\eta_{ii})^2 - \det(\boldsymbol{\eta}), \tag{74}$$

and

$$\chi_{ij} = \mathcal{B}_{jilk}\eta_{kl} + \frac{1}{2}\mathcal{B}^2_{jilknm}\eta_{kl}\eta_{mn} + p(\eta_{ji} - \eta_{jk}\eta_{ki}) - p^*(\delta_{ji} - \eta_{ji}) + \cdots, \tag{75}$$

where p and $p + p^*$ are the Lagrange multipliers enforcing incompressibility in B_e and B_t, respectively, and $\mathcal{B}_{jilk}$ and $\mathcal{B}^2_{jilknm}$ are the first- and second-order instantaneous elastic moduli whose expressions in terms of principal stretches in the primary deformation can be found in Ogden (1984) or Fu and Ogden (1999).

We now specialise to a sufficiently large circular hyperelastic plate where the deformation from B_0 to B_e corresponds to an all-round tension in its plane. The plate thicknesses in B_0 and B_e are denoted by H and h, respectively. We shall focus on axisymmetric deformations, and so cylindrical coordinates will be employed, with coordinates θ, z and r in B_e corresponding to 1-, 2-, and 3-directions, respectively. The r and θ coordinates define positions in the plane, whereas z measures distance in the out-of-plane direction with $z = 0$ corresponding to the mid-plane and the surfaces to $z = \pm h/2$. Thus, corresponding to the equibiaxial deformation, the three principal stretches are given by

$$\lambda_1 = \lambda_3 = \lambda, \quad \lambda_2 = \lambda^{-2} \tag{76}$$

in terms of a single parameter λ due to the constraint of incompressibility.

We decompose the incremental displacement $\boldsymbol{u}$ in the form

$$\boldsymbol{u} = u(r, z)\boldsymbol{e}_r + v(r, z)\boldsymbol{e}_z, \tag{77}$$

where $\boldsymbol{e}_r$ and $\boldsymbol{e}_z$ are the basis vectors in the r- and z-directions, respectively, and u and v are the associated displacement components. It then follows that

$$\eta = \text{grad}\,\boldsymbol{u} = \frac{u}{r}\boldsymbol{e}_\theta \otimes \boldsymbol{e}_\theta + v_z\boldsymbol{e}_z \otimes \boldsymbol{e}_z + v_r\boldsymbol{e}_z \otimes \boldsymbol{e}_r + u_z\boldsymbol{e}_r \otimes \boldsymbol{e}_z + u_r\boldsymbol{e}_r \otimes \boldsymbol{e}_r, \tag{78}$$

where $v_z = \partial v/\partial z$, $v_r = \partial v/\partial r$, etc. For the current axisymmetric deformation, only the equilibrium equations corresponding to $i = 2, 3$ are not satisfied automatically, and they are given by

$$\chi_{3j,j} + \frac{1}{r}(\chi_{33} - \chi_{11}) = 0, \quad \chi_{2j,j} + \frac{1}{r}\chi_{23} = 0. \tag{79}$$

We assume that both the top and bottom surfaces are traction-free, so the associated incremental boundary conditions are

$$\chi_{22} = 0, \quad \chi_{32} = 0, \quad \text{on } z = \pm h/2. \tag{80}$$

4.2 *Linear Analysis*

The linearised incompressibility condition takes the form

$$\frac{\partial (ru)}{\partial r} + \frac{\partial (rv)}{\partial z} = 0. \tag{81}$$

This is to be solved together with the equilibrium equations (79) subject to the boundary conditions (80) with the non-zero stress components given by

$$\chi_{11} = (\mathcal{B}_{1111} + p)\frac{u}{r} + \mathcal{B}_{1122}v_z + \mathcal{B}_{1133}u_r - p^*, \tag{82}$$

$$\chi_{22} = \mathcal{B}_{1122}\frac{u}{r} + (\mathcal{B}_{2222} + p)v_z + \mathcal{B}_{2233}u_r - p^*, \tag{83}$$

$$\chi_{33} = \mathcal{B}_{1133}\frac{u}{r} + \mathcal{B}_{2233}v_z + (\mathcal{B}_{3333} + p)u_r - p^*, \tag{84}$$

$$\chi_{23} = \mathcal{B}_{3232}v_r + (\mathcal{B}_{3223} + p)u_z, \tag{85}$$

$$\chi_{32} = \mathcal{B}_{2323}u_z + (\mathcal{B}_{2332} + p)v_r. \tag{86}$$

Equation (81) can be satisfied by introducing a 'stream function' $\phi(r, z)$ such that

$$u = \frac{1}{r}\phi_z, \quad v = -\frac{1}{r}\phi_r, \tag{87}$$

where as in (78) a subscript signifies differentiation (e.g. $\phi_z = \partial\phi/\partial z$). As a result, the above eigenvalue problem may be reduced to an eigenvalue problem for ϕ that admits a "normal mode" buckling solution of the form

$$\phi(r, z) = rJ_1(kr)S(kz), \tag{88}$$

where k is a constant playing the role of wavenumber, $J_1(x)$ is the Bessel function of the first kind, and the other function $S(kz)$ is to be determined. This is similar to looking for periodic bifurcation solutions proportional to $\mathrm{e}^{\mathrm{i}kx_1}$ in rectangular coordinates. In fact, $J_1(x)$ is oscillatory although it also decays like $1/\sqrt{x}$ for large values of x due to geometric spreading.

Due to the symmetry of the geometry of the plate and boundary conditions with respect to the mid-plane $z = 0$, this eigenvalue problem admits extensional and flexural modes that are decoupled from each other. It can be shown that the bifurcation condition for the extensional modes is given by

$$\sqrt{s_1}\,(1 + s_1)\,(s_2(2\beta + \gamma) - \gamma)\tanh\left(\frac{kh}{2\sqrt{s_1}}\right)$$

$$- \sqrt{s_2}\,(1 + s_2)\,(s_1(2\beta + \gamma) - \gamma)\tanh\left(\frac{kh}{2\sqrt{s_2}}\right) = 0, \tag{89}$$

where

$$\alpha = \mathcal{B}_{1212}, \quad 2\beta = \mathcal{B}_{1111} + \mathcal{B}_{2222} - 2\mathcal{B}_{1122} - 2\mathcal{B}_{1221}, \quad \gamma = \mathcal{B}_{2121}, \tag{90}$$

and

$$s_1 = \frac{1}{\alpha}(\beta - \sqrt{\beta^2 - \alpha\gamma}), \quad s_2 = \frac{1}{\alpha}(\beta + \sqrt{\beta^2 - \alpha\gamma}). \tag{91}$$

We refer to Wang et al. (2022) for details of the derivations. Expanding (89) to order $(kh)^2$, we obtain

$$\gamma(\beta+\gamma)+\frac{1}{24}\left\{\alpha\gamma-(2\beta+\gamma)^2\right\}(kh)^2+\cdots=0, \tag{92}$$

where we have used (91) to eliminate s_1 and s_2.

From the discussion in Fu (2001), we may postulate that the bifurcation condition for localized necking can be obtained by setting the leading order term in (92) to zero, that is $\beta+\gamma=0$ since $\gamma>0$, or equivalently,

$$\mathcal{B}_{1111}+\mathcal{B}_{2222}+2\mathcal{B}_{2121}-2\mathcal{B}_{1221}-2\mathcal{B}_{1122}=0. \tag{93}$$

In terms of the strain-energy function $W(\lambda_1,\lambda_2,\lambda_3)$, this bifurcation condition takes the form

$$2\lambda^2\, W_2^{(0)}+W_{22}^{(0)}-2\lambda^3 W_{12}^{(0)}+\lambda^6 W_{11}^{(0)}=0, \tag{94}$$

where $W_2=\partial W/\partial\lambda_2$, $W_{12}=\partial^2 W/\partial\lambda_1\partial\lambda_2$ etc., and a superscript "(0)" signifies evaluation at $(\lambda_1,\lambda_2,\lambda_3)=(\lambda,\lambda^{-2},\lambda)$.

The above bifurcation condition has the following interpretations. The nominal stresses in the x_3- and x_2-directions (i.e. r- and z-directions) are given by

$$S_3=W_3-\lambda_3^{-1}p, \quad S_2=W_2-\lambda_2^{-1}p.$$

On eliminating p using the condition $S_2=0$, we obtain $p=\lambda_2 W_2$ and

$$S_3=W_3-\frac{\lambda_2}{\lambda_3}W_2=\frac{\partial\tilde{W}}{\partial\lambda_3},$$

where $\tilde{W}$ is the reduced strain energy function defined by

$$\tilde{W}(\lambda_1,\lambda_3)=W(\lambda_1,\lambda_1^{-1}\lambda_3^{-1},\lambda_3). \tag{95}$$

Similarly, we have $S_1=\partial\tilde{W}/\partial\lambda_1$. From now on we view S_1 and S_3 as functions of λ_1 and λ_3. It can then be shown that the bifurcation condition (93) is equivalent to

$$\left.\frac{\partial S_3}{\partial\lambda_3}\right|_{\lambda_3=\lambda_1=\lambda}=0. \tag{96}$$

In the case of localized bulging of an inflated hyperelastic tube, the bifurcation condition was shown to be equivalent to the Jacobian determinant of the internal pressure P and resultant axial force N equal to zero when P and N are each viewed as functions of two deformation variables such as the internal volume and axial stretch (Fu et al., 2016). In the current setting, it is natural to choose the nominal stresses S_1 and S_3 as the force variables and λ_1 and λ_3 as the deformation variables. The associated Jacobian determinant can be evaluated at $\lambda_1=\lambda_3=\lambda$ as follows:

$$\frac{\partial S_1}{\partial \lambda_1}\frac{\partial S_3}{\partial \lambda_3} - \frac{\partial S_1}{\partial \lambda_3}\frac{\partial S_3}{\partial \lambda_1} = \left(\frac{\partial S_3}{\partial \lambda_3}\right)^2 - \left(\frac{\partial S_3}{\partial \lambda_1}\right)^2 = \left(\frac{\partial S_3}{\partial \lambda_3} + \frac{\partial S_3}{\partial \lambda_1}\right)\left(\frac{\partial S_3}{\partial \lambda_3} - \frac{\partial S_3}{\partial \lambda_1}\right), \tag{97}$$

where all the partial derivatives are evaluated at $\lambda_1 = \lambda_3 = \lambda$ and so indexes 1 and 3 can be exchanged. Thus, the Jacobian determinant equal to zero implies

$$\frac{\partial S_3}{\partial \lambda_3} + \frac{\partial S_3}{\partial \lambda_1} = 0, \quad \text{or} \quad \frac{\partial S_3}{\partial \lambda_3} - \frac{\partial S_3}{\partial \lambda_1} = 0, \tag{98}$$

which may be compared with the condition (96) for necking. The two equations in (98) are obviously equivalent to

$$\frac{d}{d\lambda} S_3(\lambda, \lambda) = 0, \quad \text{or} \quad \left\{ \lim_{\lambda_3 \to \lambda_1} \frac{S_3 - S_1}{\lambda_3 - \lambda_1} \right\}_{\lambda_1 = \lambda} = 0. \tag{99}$$

The first condition corresponds to the limit point instability. The second condition in $(99)_2$ corresponds to $S_3 = S_1$ with the trivial solution $\lambda_3 = \lambda_1$ factored out. In other words, it is the condition when $S_3 = S_1$ starts to be satisfied by a deformation with $\lambda_3 \neq \lambda_1$. This is nowadays known as the Treloar-Kearsley (TK) instability due to Treloar (1947); Kearsley (1986) although its plane-strain counterpart was studied earlier by Ogden (1985); see also Ogden (1987).

Thus, for the current problem, *the Jacobian determinant equal to zero is not the bifurcation condition for necking.*

For the purpose of illustration, we consider the following class of strain-energy functions:

$$W = \frac{2\mu_1}{m_1^2}(\lambda_1^{m_1} + \lambda_2^{m_1} + \lambda_3^{m_1} - 3) + \frac{2\mu_2}{{m_2}^2}(\lambda_1^{m_2} + \lambda_2^{m_2} + \lambda_3^{m_2} - 3), \tag{100}$$

where μ_1, μ_2, m_1, m_2 are material constants. We then have

$$\left\{\frac{\partial S_1}{\partial \lambda_3}\right\}_{\lambda_1=\lambda_3=\lambda} = \left\{\frac{\partial S_3}{\partial \lambda_1}\right\}_{\lambda_1=\lambda_3=\lambda} = \frac{2}{\lambda^2}\left(\mu_1 \lambda^{-2m_1} + \mu_2 \lambda^{-2m_2}\right), \tag{101}$$

which is positive provided both μ_1 and μ_2 are positive. Assuming that the latter is true, it then follows from (98) that necking would occur after the TK instability but before the limit point instability. For instance, for the case when $\mu_2 = 0,\ m_1 = 1/2$, we have

$$\lambda_{\text{TK}} = 2^{2/3}, \quad \lambda_{\text{necking}} = 3^{2/3}, \quad \lambda_{\text{limit}} = 4^{2/3}, \tag{102}$$

where the three subscripts in λ signify association with the TK, necking, and limit point instabilities, respectively. On the other hand, when $\mu_2 = \mu_1/12,\ m_1 = 1/2,\ m_2 = 2$ with which the material exhibits moderate stiffening for large values of λ, we have

$$\lambda_{\rm TK1}=1.698,\quad \lambda_{\rm TK2}=7.803,\quad \lambda_{\rm necking1}=2.318,$$

$$\lambda_{\rm necking2}=7.487,\quad \lambda_{\rm limit1}=2.957,\quad \lambda_{\rm limit2}=7.102,$$

where we have two bifurcation values for each of the three instabilities. To be more precise, the second bifurcation value for necking is really the value for *bulging* when unloading takes place. We expect, however, that the TK instability can only occur under force control even if the bifurcation condition $(98)_1$ is satisfied. Under displacement control, the two in-plane principal stretches are forced to be equal, and we expect that the TK instability should give way to necking if (94) is satisfied.

4.3 Weakly Nonlinear Analysis

Having determined the bifurcation condition for necking, we now denote the critical stretch by $\lambda_{\rm cr}$ and conduct a weakly nonlinear analysis to derive an amplitude equation for the necking solution. To simplify notations, we scale r, z and all displacement components by h, and use the same notations for the scaled quantities. As a result, we have $h = 1$.

We use $\lambda_1 = \lambda_3 \equiv \lambda$ as the control parameter in our near-critical analysis. From the linear analysis we deduce that in a small neighbourhood of $\lambda = \lambda_{\rm cr}$, we have $k^2 \sim (\lambda - \lambda_{\rm cr})$,

$$u \sim kJ_1(kr),\qquad v \sim -\frac{1}{kr}\left\{J_1(kr) + krJ_1'(kr)\right\}k^2.$$

Thus, $\lambda - \lambda_{\rm cr} = O(k^2)$, $v = O(ku)$, $p^* = O(ku)$ (we have not written out the expression for p^* for brevity), and dependence on r is through kr. Thus, we write

$$\lambda = \lambda_{\rm cr} + \epsilon\lambda_0, \tag{103}$$

and define a far distance variable s through

$$s = \sqrt{\epsilon}r, \tag{104}$$

where λ_0 is a constant and ϵ is a small positive parameter characterizing the order of deviation of λ from its critical value $\lambda_{\rm cr}$. Note that as a result of (103), the moduli $\mathcal{B}_{jilk}$ and B^2_{jilknm} must all be expanded in terms of ϵ as well, but these expansions are not written out for the sake of brevity. The pressure p is eliminated using the identity $p = \mathcal{B}_{2121} - \mathcal{B}_{2112}$.

It can be shown that u must be of order $\sqrt{\epsilon}$; see, e.g., Fu (2001). Thus, we look for an asymptotic solution of the form

$$u = \sqrt{\epsilon}\left\{u^{(1)}(s,z) + \epsilon u^{(2)}(s,z) + \epsilon^2 u^{(3)}(s,z) + \cdots\right\},$$

$$v = \epsilon\left\{v^{(1)}(s,z) + \epsilon v^{(2)}(s,z) + \epsilon^2 v^{(3)}(s,z) + \cdots\right\}, \tag{105}$$

$$p^* = \epsilon\left\{p^{(1)}(s,z) + \epsilon p^{(2)}(s,z) + \epsilon^2 p^{(3)}(s,z) + \cdots\right\},$$

where all the functions on the right hand sides are to be determined from successive approximations.

On substituting (105) into the equilibrium equations (79), the incompressibility condition (74), the boundary conditions (80), and then equating the coefficients of like powers of ϵ, we obtain a hierarchy of boundary value problems. Solving the leading order problem, we obtain

$$u^{(1)}(s,z) = A(s), \quad v^{(1)}(s,z) = -z\frac{1}{s}(sA(s))' + B(s), \tag{106}$$

where $A(s)$ and $B(s)$ are functions to be determined at higher orders. Solving the above boundary value problems at higher orders requires the satisfaction of a series of solvability conditions. The condition obtained at second order is simply the bifurcation condition (93). The solvability condition at the third order takes the form of an amplitude equation for $A(s)$:

$$c_0\frac{d}{ds}\frac{1}{s}\frac{d}{ds}sP'(s) + c_1\lambda_0 P'(s) + c_2\frac{d}{ds}P^2(s)$$
$$+ c_3 A''(s)\left(A'(s) - \frac{1}{s}A(s)\right) = 0, \tag{107}$$

where a prime signifies differentiation, $P(s)$ is defined by

$$P(s) = \frac{1}{s}(sA(s))', \tag{108}$$

and the four coefficients are given by

$$\begin{aligned}
c_0 &= \frac{1}{12}\left(\mathcal{B}_{1212} - \mathcal{B}_{2121}\right),\\
c_1 &= 2\mathcal{B}'_{1122} + 2\mathcal{B}'_{1221} - \mathcal{B}'_{1111} - 2\mathcal{B}'_{2121} - \mathcal{B}'_{2222},\\
c_2 &= \frac{1}{4}\left(-4\mathcal{B}_{1111} - 2\mathcal{B}_{1122} + 6\mathcal{B}_{2222} - \mathcal{B}^2_{111111} + 4\mathcal{B}^2_{111122} - \mathcal{B}^2_{111133}\right.\\
&\quad \left.-6\mathcal{B}^2_{112222} + 2\mathcal{B}^2_{112233} + 2\mathcal{B}^2_{222222}\right),\\
c_3 &= \mathcal{B}_{1122} - \mathcal{B}_{1111} + \mathcal{B}^2_{111122} - \mathcal{B}^2_{112233} - \frac{1}{2}\mathcal{B}^2_{111111} + \frac{1}{2}\mathcal{B}^2_{111133}.
\end{aligned}$$

In the above expressions, $\mathcal{B}'_{1122}$ denotes $d\mathcal{B}_{1122}/d\lambda$ etc., and all the elastic moduli are evaluated at $\lambda = \lambda_{\rm cr}$.

As a consistency check, we may expand (108) fully out and omit all the terms that are divided by powers of s to obtain its planar counterpart:

$$c_0 A^{(4)}(s) + c_1\lambda_0 A''(s) + c_2^* A'(s)A''(s) = 0, \tag{109}$$

where

$$c_2^* = 2c_2 + c_3 = 3\mathcal{B}_{2222} - 3\mathcal{B}_{1111} - \mathcal{B}^2_{111111} + 3\mathcal{B}^2_{111122} - 3\mathcal{B}^2_{112222} + \mathcal{B}^2_{222222}. \tag{110}$$

Equation (109) applies to a finitely deformed plate that is subjected to plane-strain incremental deformations, and recovers the static counterpart of the amplitude equation (4.41) in Fu (2001). It has an exact solution given by

$$A(s) = \frac{6c_0}{c_2^*}\sqrt{\frac{-c_1\lambda_0}{c_0}}\tanh\left(\frac{1}{2}\sqrt{\frac{-c_1\lambda_0}{c_0}}s\right). \tag{111}$$

This solution has the property $A'(s) \to 0$ as $s \to \infty$ and is the localized necking solution in the 2D case.

Although the solution (111) for the plane-strain case tends to a constant as $s \to \infty$, we expect its counterpart for the axisymmetric case to decay algebraically due to geometrical spreading. Thus, for large s, we may neglect the quadratic part in (107) to obtain

$$\frac{d}{ds}\frac{1}{s}\frac{d}{ds}sP'(s) - \kappa^2 P'(s) = 0, \tag{112}$$

where $\kappa^2 = -c_1\lambda_0/c_0$. Integrating once and setting the constant to zero, we obtain

$$s^2 P''(s) + sP'(s) - \kappa^2 s^2 P(s) = 0.$$

This is a modified Bessel's equation. It has a decaying solution given by

$$P(s) = a_1 K_0(\kappa s), \tag{113}$$

where K_0 is the modified Bessel function of the second kind and a_1 is a constant. On substituting (113) into (108) and integrating, we obtain

$$A(s) = A_\infty(s) \equiv \frac{a_2}{s} - \frac{a_1}{\kappa}K_1(\kappa s), \tag{114}$$

where a_1 and a_2 are constants, and K_1 is the modified Bessel function of the second kind that has the asymptotic behaviour

$$K_1(x) \sim \sqrt{\frac{\pi}{2x}}\left[1 + \frac{3}{8x} + \cdots\right]e^{-x}, \quad \text{as } x \to \infty. \tag{115}$$

The asymptotic behaviour (114) is consistent with our earlier assumption that $A(s)$ decays algebraically.

Thus, the 4th-order differential equation (107) can be integrated subject to the conditions

$$A(s),\, A''(s) \to 0 \;\; \text{as } s \to 0, \tag{116}$$

and

$$A(s) \to A_\infty(s), \;\; A'(s) \to A'_\infty(s) \;\; \text{as } s \to \infty. \tag{117}$$

This is carried out using the central finite difference method as follows. We replace $[0, \infty)$ by $[0, L]$ where L is a sufficiently large positive number. As in Sect. 3.4, we discretize the interval $[0, L]$ into n intervals with node points $s_i = iL/n$ ($i = 0, 1, \ldots, n$). At the i-th node, the first four derivatives are replaced by

$$A' = \frac{A_{i+1} - A_{i-1}}{2h}, \qquad A'' = \frac{A_{i+1} - 2A_i + A_{i-1}}{h^2},$$

$$A''' = \frac{A_{i+2} - 2A_{i+1} + 2A_{i-1} - A_{i-2}}{2h^3},$$

$$A^{(4)} = \frac{A_{i+2} - 4A_{i+1} + 6A_i - 4A_{i-1} + A_{i-2}}{h^4}, \tag{118}$$

where

$$A_i \equiv r(s_i) \quad h = \frac{L}{n}.$$

Applying this scheme at the node points $Z_1, Z_2, \ldots, Z_n$, we obtain n nonlinear homogeneous equations for the n unknowns $A_1, \ldots, A_n$, noting that $A_0 = 0$ according to (116). These equations also involve A_{-1}, A_{n+1} and A_{n+2}, which are determined as follows. First, the A_{-1} is determined using

$$A''(0) = \frac{A_1 - 2A_0 + A_{-1}}{h^2} = 0, \quad \Longrightarrow \quad A_{-1} = -A_1.$$

To determine A_{n+1} and A_{n+2}, we use an idea that is often used in imposing asymptotic boundary conditions. Suppose as $s \to \infty$ a function $y(s)$ decays like $y(s) = b\mathrm{e}^{-as}$ for some constants $a (> 0)$ and b. It then follows that $y'(s) + ay(s)$ decays to zero faster than e^{-as}, which means that with the same accuracy $y'(L) + ay(L) = 0$ can be imposed at a smaller value of L than when $y(L) = 0$ is imposed. Extending this idea, we use (114) to compute

$$A'(s) = -\frac{a_2}{s^2} - a_1 K_1'(\kappa s). \tag{119}$$

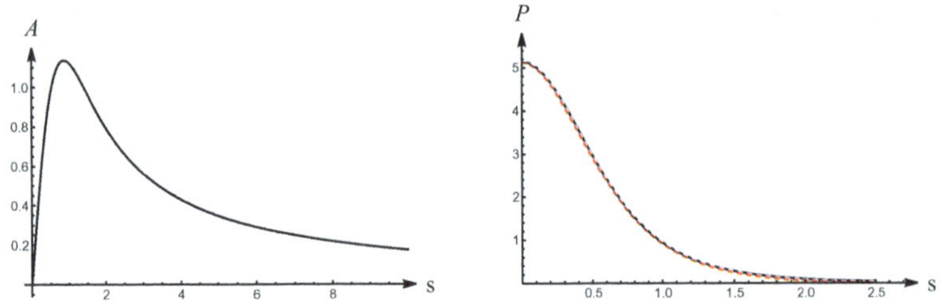

(a) Profile of A (b) Profile of P

Fig. 19 Finite difference solutions of the amplitude equation (107). In (b), the black and red lines are the FD solution and the solution given by (121), respectively

On evaluating (114) and (119) at $s = s_n$ and solving the two resulting equations simultaneously, we may express a_1 and a_2 in terms of $A(s_n)$ and $A'(s_n)$. We impose the asymptotic boundary conditions

$$A''(s_n) = A''_\infty(s_n), \quad A'''(s_n) = A'''_\infty(s_n). \tag{120}$$

The two constants a_1 and a_2 in $A_\infty(s)$ have just been expressed in terms of $A(s_n)$ and $A'(s_n)$. The discretized forms of (120) then yield two equations for A_{n+1} and A_{n+2}.

The above system of n nonlinear equations is solved using the planar solution (111) multiplied by $1/(1+s)$ as an initial guess although other initial guesses can also be used (Fu and Yu, 2023). As an example, we consider the case when the strain energy function is given by (100) with $\mu_2 = 0,\ m_1 = 1/2$. We then have

$$c_1/c_0 = 9\sqrt[3]{3}/2, \quad c_2/c_0 = 27/8, \quad c_3/c_0 = 9/4.$$

Figures 19a, b show our FD solutions for A and P, respectively. In Fig. 19b we have also shown in red dashed line the function

$$P(s) = \frac{a}{bs^2+1}\text{sech}^2(cs), \tag{121}$$

where $a = 5.27,\ b = -0.11,\ c = 1.45$. These constants are obtained by fitting the approximate representation (121) to the FD solution for P.

4.4 Fully Nonlinear Regime

Abaqus simulations of the current axisymmetric necking problem have been carried out in Wang et al. (2022). It was shown that axisymmetric necking follows a similar evolution process as localized bulging of inflated hyperelastic tubes: as the imposed

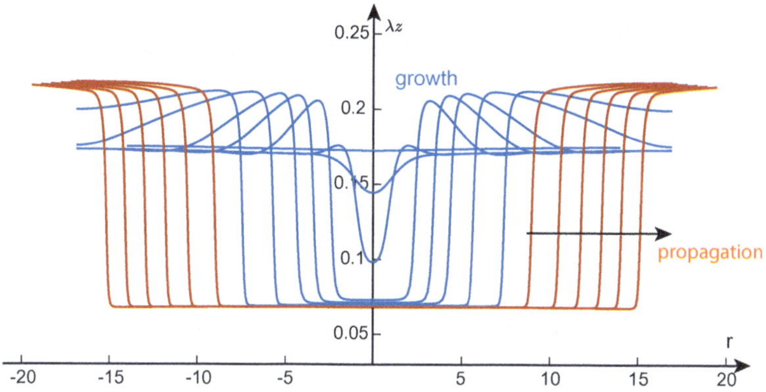

Fig. 20 Initiation, growth, and propagation of necking in a circular plate under all-round stretching. From Wang et al. (2022)

edge displacement in the radial direction is increased gradually, the necking initiates, grows, and then propagates in the radial direction; see Fig. 20. In the propagation stage, the circular plate consists of two deformation "phases": a circular region centred around the origin where the deformation is homogeneous and an annular region where the deformation is inhomogeneous. The two regions are connected by a thin transition region just as in the inflation problem.

5 Conclusion

In this chapter, we have discussed three elastic localization problems within the framework of nonlinear elasticity. Although much less is previously known for such localization problems than for problems involving periodic modes, the entire localization process can in fact be understood analytically or sem-analytically using the methodology explained in this chapter. To make the problem tractable analytically, especially in the fully nonlinear regime, we have assumed, for instance in the first problem, the tube length to be infinite. Although very few objects are truly infinite in any direction, it is shown that for localization analysis it is best to treat the dimension in the direction of localization as infinite unless the actual dimension for a particular application is so small that localization is no longer observable. Another important feature is that 1D gradient models are expected to play an important role in future studies. Such a 1D model has been derived by Lestringant and Audoly (2018, 2020) for the inflation problem under the membrane assumption and for the problem of surface tension-induced necking, respectively. The corresponding 1D model for the inflation problem associated with a tube of finite wall thickness has recently been derived by Yu and Fu (2023), and our recent studies have shown that a 1D model for the axisymmetric necking problem can also be derived. All these ideas can be extended to other similar localization problems, in particular to problems that involve other or multiple fields such as electric/magnetic fields, chemical reactions, residual stresses, internal damage, and even to materials experiencing plastic deformations as long as unloading is not involved.

References

Alexander, H. (1971). Tensile instability of initially spherical balloons. *International Journal of Engineering Science, 9*(1), 151–160.

Chater, E., & Hutchinson, J. W. (1984). On the propagation of bulges and buckles. *ASME Journal of Applied Mechanics, 51*, 269–277.

Emery, D., & Fu, Y. B. (2021). Localised bifurcation in soft cylindrical tubes under axial stretching and surface tension. *International Journal of Solids and Structures, 219–220*, 23–33.

Emery, D., & Fu, Y. B. (2021). Post-bifurcation behaviour of elasto-capillary necking and bulging in soft tubes. *Proceedings of the Royal Society A, 477*, 20210311.

Fu, Y. B. (2001). Nonlinear stability analysis. In Y. B. Fu, R. W. Ogden (Eds.), *Nonlinear elasticity: Theory and applications*. Cambridge: Cambridge University Press.

Fu, Y. B., & Ogden, R. W. (1999). Nonlinear stability analysis of pre-stressed elastic bodies. *Continuum Mechanics and Thermodynamics, 11*, 141–172.

Fu, Y. B., & Yu, X. (2023). Axisymmetric necking of a circular electrodes-coated dielectric membrane. *Mechanics of Materials, 181*, 104645.

Fu, Y. B., Liu, J. L., & Francisco, G. S. (2016). Localized bulging in an inflated cylindrical tube of arbitrary thickness—The effect of bending stiffness. *Journal of the Mechanics and Physics of Solids, 90*, 45–60.

Fu, Y. B., Jin, L. S., & Goriely, A. (2021). Necking, beading, and bulging in soft elastic cylinders. *Journal of the Mechanics and Physics of Solids, 147*, 104250.

Guo, Z. M., Wang, S. B., & Fu, Y. B. (2022). Localised bulging of an inflated rubber tube with fixed ends. *Proceedings of the Royal Society A, 380*, 20210318.

Kearsley, E. A. (1986). Asymmetric stretching of a symmetrically loaded elastic sheet. *International Journal of Solids and Structures, 22*, 111–119.

Kirchgässner, K. (1982). Wave solutions of reversible systems and applications. *Journal of Differential Equations, 45*, 113–127.

Lestringant, C., & Audoly, B. (2018). A diffuse interface model for the analysis of propagating bulges in cylindrical balloons. *Proceedings of the Royal Society A, 474*, 20180333.

Lestringant, C., & Audoly, B. (2020). A one-dimensional model for elasto-capillary necking. *Proceedings of the Royal Society A, 476*, 20200337.

Ogden, R. W. (1984). *Non-linear elastic deformations*. New York: Ellis Horwood.

Ogden, R. W. (1985). Local and global bifurcation phenomena in plane-strain finite elasticity. *International Journal of Solids and Structures, 21*, 121–132.

Ogden, R. W. (1987). On the stability of asymmetric deformations of a symmetrically tensioned elastic sheet. *International Journal of Engineering Science, 25*, 1305–1314.

Taffetani, M., & Ciarletta, P. (2015). Beading instability in soft cylindrical gels with capillary energy: Weakly non-linear analysis and numerical simulations. *Journal of the Mechanics and Physics of Solids, 81*, 91–120.

Treloar, L. R. G. (1947). Stresses and birefringence in rubber subjected to general homogeneous strain. *Proceedings of the Physical Society A, 60*, 135–145.

Wang, M., Jin, L. S., & Fu, Y. B. (2022). Axi-symmetric necking versus Treloar-Kearsley instability in a hyperelastic sheet under equibiaxial stretching. *Mathematics and Mechanics of Solids, 27*, 1610–1631.

Wang, S. B., Guo, Z. M., Zhou, L., Li, L. A., & Fu, Y. B. (2019). An experimental study of localized bulging in inflated cylindrical tubes guided by newly emerged analytical results. *Journal of the Mechanics and Physics of Solids, 124*, 536–554.
Wilkes, E. W. (1955). On the stability of a circular tube under end thrust. *Quarterly Journal of Mechanics and Applied Mathematics, 8*(1), 88–100.
Xuan, C., & Biggins, J. (2017). Plateau-rayleigh instability in solids is a simple phase separation. *Physical Review E, 95*, 053106.
Ye, Y., Liu, Y., & Fu, Y. B. (2020). Post-bifurcation analysis of a thin-walled hyperelastic tube under inflation. *Journal of the Mechanics and Physics of Solids, 135*, 103804.
Yin, W.-L. (1977). Non-uniform inflation of a cylindrical elastic membrane and direct determination of the strain energy function. *Journal of Elasticity, 7*, 265–282.
Yu, X., & Fu, Y. B. (2023). A one-dimensional model for axisymmetric deformations of an inflated hyperelastic tube of finite wall thickness. *Journal of the Mechanics and Physics of Solids, 175*, 105276.

SPRINGER NATURE

GPSR Compliance

The European Union's (EU) General Product Safety Regulation (GPSR) is a set of rules that requires consumer products to be safe and our obligations to ensure this.

If you have any concerns about our products, you can contact us on ProductSafety@springernature.com

In case Publisher is established outside the EU, the EU authorized representative is:

Springer Nature Customer Service Center GmbH
Europaplatz 3
69115 Heidelberg, Germany

The manufacturer's authorised representative in the EU is Springer Nature Customer Service Centre GmbH, Europaplatz 3, 69115 Heidelberg, Germany. If you have any concerns regarding our products, please contact ProductSafety@springernature.com

Printed and bound by CPI Group (UK) Ltd, Croydon, CR0 4YY
01/10/2026
02225899-0018